Sgt. Hugh A. O'Neill, Jr. - Home from War (1945)

Editor: Scott MacGregor

Library of Congress Cataloging In-Publication Data

ISBN: 979-8-9884861-0-7
ISBN: 979-8-9884861-1-4
eISBN: 979-8-9884861-2-1

Library of Congress Control Number: 2023944905

Text Copyright ©2023 Hugh A. O'Neill, Jr.
Story Illustrations Copyright ©2023 Gary Dumm
Front and Back Cover Illustrations: Gary Dumm
Additional Text Copyright 2023 Scott MacGregor

Book Design and Gray-Scale Colorization of Artwork: Scott MacGregor

THIS BOOK IS A 100% HUMAN CREATION!
<u>NO</u> ARTIFICIAL INTELLIGENCE WAS USED
IN ITS PRODUCTION

Cleveland, Ohio

Published in 2023 by EOI Media Press, Inc. All rights reserved. No portion of this book may be reproduced, stored in a retrieval system or transmitted in any form or by any means, mecahancal, electronic, photocopying, recording, or otherwise, without written permission from the publisher.

CONTENTS

Foreward by Scott MacGregor..I
Preface by Hugh O'Neill...III

Part 1: Captured!..1
Part 2: The Shutters of Consciousness...12
Part 3: An Affirmation of Spirit..22
Part 4: The Stench of Fear...37
Part 5: The Inhalations of Air..51
Part 7: "Schweinhundts"...74
Part 8: The Writhing Maw of the Monster......................................89
Part 9: That Unforgettable Place...97
Part 10: Like Dante's Inferno...110
Part 11: Coda..120

The Post-War Years by Scott MacGregor..VI
Bibliography..XVI
About the Author, Illustrator, and Editor..XVII

List of Illustrations (by Gary Dumm)

Captured!..3
Killed By Our Own Artillery..7
Men Gone Mad...17
Filthy Hands...19
Our Daily Bread...25
Soothed By Song..33
Soup Good and Hot...41
A Terrifying Light..47
American Hitchhiker...55
The Unwanted Gift..61
The Unsuspecting Soldier...65
A Roadside Cremation..67
Death From the Sky..71
The Pie-Eaters...77
Leaving the Earth..81
A Red Cross Nazi..83
The Storybook Village..87
The Potato Farmer..95
A Rifle Butt to the Head..101
Damn the Skies..105
Have a Cigar...107
Dangling Like Monkeys..115
Countryside Sadists...117
Blue Skies and Geraniums..123

*Dedicated to the Memory
of Seon O'Neill
1947-2017*

Foreword

Hugh O'Neill, the author of this memoir, was my uncle. He was my mother's younger and only sibling. It has been over twenty years since his passing, and his family, including myself remains very proud of him for the service he gave to our country.

It's believed that Hugh was in his early 30's when he penned this memoir. It's very existence was unknown to his family until after his passing in 2001 at the age of 80. Readers will soon discover that Hugh O'Neill was not your typical G.I. diarist. He was, in fact, a gifted English literature and journalism student before the War, and a respected career newspaper journalist afterwards.

Sgt. Hugh A. O'Neill fought with the "Tigers" of the 61st Armored Infantry Battalion, an element of the 10th Armored Division. Drafted in 1942 at the age of 21, he'd underwent his basic training at Fort Benning, Georgia, and shipped out to face action on European battlefields in August 1944.

The 10th Armored Division was a storied component of the American Third Army, led by General George S. Patton. Under Patton's command, Third Army had spent the autumn of 1944 pushing the German Army out of France, Luxembourg, and Belgium. When the surprise German Winter Offensive of December 1944 (aka: "The Battle of the Bulge") had begun, Hugh O'Neill was in a front row seat aboard an armored halftrack. The subsequent "Siege of Bastogne" is known, (and rightly so), as a great victory for the American 101st Airborne Division. It's a forgotten fact that when the Germans broke through the Allied lines, the 10th Armored Division stood fast and held onto the strategic Belgian roads around

Bastogne until the Airborne arrived. Hugh had survived the siege of Bastogne, but a month later, while on a counter-reconaissance mission along the German border, he and others had sought shelter from the winter cold inside a seemingly abandoned German bunker. Once inside, the Germans returned and ruthlessly attacked the fortification, killing 20 Americans. Sgt. O'Neill was dragged out unconscious, and alive, but for how long? By 1945, German prison camps were overflowing and POWs were a huge inconvenience for the decimated German military. Still fresh in mind was the murder of 84 unarmed American POWs' by the German SS, only a few weeks earlier. Instead of a firing squad, Hugh and his fellow captives were forced on a four month long "Death March" across the roads of southern Germany in the dead of winter without proper clothing, food, or shelter.

There are those who believe that a captured soldier is a "failed soldier". Hugh himself ponders this question in his memoir, as was his personal right, despite fighting bravely in several of the War's most consequential battles. His remarkable story of survival is yet another cautionary tale from the world's greatest conflict.

Whether we choose to understand it or not, the consequences of both World Wars will continue to affect and inform our present and future, possibly forever. First person accounts of war's horrors will always be useful so that future generations are provided with real world context and insight, instead of the videogame variety. Let's pray they pay attention. Otherwise, war may continue to be, as O'Neill calls it: "the destiny of our kind."

- Scott MacGregor, 2023

Preface

The incidents of war are many, and because all are unpleasant, they are soon forgotten. We never realize that war is upon us until the first man dies. With the passing of the last we try to forget that the ugly years of conflict ever existed. In this way we tolerate a war, answer its demands, and never anticipate another. Yet man has entered some 900 wars in 2500 years and will participate in many more before the earth reaches its end times and tosses its humans like so many pebbles into the void.

All wars stand in an unbroken line of succession be they so called world wars, civil wars, just or unjust wars. War is war and human beings' fight. Whether from club to cannon, or flame thrower to atomic bomb, there are no differences. Men fight and men die and that is the basic and horrible reality of the condition.

Even though we do not like to be reminded of the wars we have known, we cannot disregard the fact that the devastations of war were always known to us. Few understand the full implications of war. Fewer possess even a partial understanding. If every person had the endowment of full vision and saw what the actuality of war is, then the world would change, and there would be no more war.

Sadly, that would be a world of people very different than those who walk our streets. It would be a world in which economic aggression would not be the prelude to overt, death-dealing aggression. In other words, it would not be our world.

Perhaps, for now, the most we can hope for is that we come to a broader understanding of how war affects the men and women who fought it. If we can grasp those realities, even slightly, we may discover that it's possible for people to walk straight, motivated by the proper proportion of joy and pride.

For my part, only this: In World War II, I was one of millions actively fighting. Ours was the European Theatre of Operations. I fought for a time in the usual pattern of hoping to live as I watched death all around me. Then, and four months before the fighting stopped, I was taken prisoner by the German army in January 1945. For me it remained the condition of war, though circumstances had altered so much that the whole conduct of the war had changed. Looking back at that life while a prisoner of war is viewing only a very small part of the activity we call war. What follows is an attempt to see those months honestly, not as anything exceptional or carried along by honor or high tragedy. Elaboration is a luxury, and war, being what it is, should not be offered amenities.

Today, this seems as though it is a view into another epoch, as indeed it is. The pattern of war, as I knew it, won't be repeated. Hiroshima and Nagasaki have changed that pattern and moved humanity into a new and ever more tenuous relationship with the Earth. The one thing we never seem to alter is death. Whether by gunpowder or Strontium 90, the death count will continue because of people who defend war's cruelty with the everlasting lie that it's the destiny of our kind. - *Hugh O'Neill*

Captured! A World War II Memoir

"The only way human beings can win a war is to prevent it."
George C. Marshall

Part I
Captured!

We had made the mistake, as events were to prove, of seeking shelter for the night in a concrete German bunker commonly known as a "pillbox". After our troops had captured the nearby village, the counter-attacking enemy savagely recaptured the bunker because its guns were pointed the right way for them and the wrong way for us.

When the Germans dragged those of us still alive out of the smashed concrete rubble, it was 2 a.m. and bitterly cold with January frost. Stars and a moon close to being full covered all of us, friends and enemies, with a neutral compassion. It was strange to realize in those first few moments that we'd been captured. Reality insisted, once we were beyond the horror of the death filled fortification, that we were free. It was to take many hours before we came to the certainty that we were prisoners.

Our captors were far from being pushed back when U.S. artillery began raining down on the open field we were crossing. The Germans delighted in the fact that Americans were being killed by their fellow Americans. It didn't matter that Germans were also being killed and wounded if the Americans were getting some of it, too. We, however, had known our guns to fire on us before in other battles, and this experience was nothing new.

Once they had marched us beyond the line of fire, we knew that we were prisoners of the enemy. The irrevocable sound of the word, the silence of the night, even the sound of the German tongue induced a weight of terror that was devoid of hope. Despair and the certainty of death was with every man. This was the first of a hundred or a thousand such instances of ambiguity. When did we realize our deaths had been thwarted and we'd been captured rather than killed? We never asked ourselves that question, and before long we had already forgotten the twenty Americans who had died in that wrecked pillbox.

Yes, despair and the certainty of death…when or how it would come we did not know, only that it would. Each sound in the night and every look that came to us from an enemy soldier brought death. In those first hours, we had died too many times to have any respect for the actuality of death, whenever it would come. We hardened our thoughts and waited under a blinding tension that consumed the small bits of energy that remained within us.

What it was or where we were taken, I do not know. My recollection is of a cave with a dark passage that led into a lighted interior. After those first few stumbling moments came vision, and with it the wariness of the trapped. The judges were on a platform, seated at a table and they were studying the charges. Unintelligible questions in what sounded like ten languages brought no answers. Our tongues were thick with fright as fingers that held our futures pointed at us. From behind the table would come a command, a soldier would move, and in the short time we inhabited the cave,

Captured!

"When the Germans dragged those of us still alive out of the smashed concrete rubble, it was 2am and cold with January frost."

our wounds were dressed. This bit of unexpected compassion brought us hope of a reprieve and changed death to life. After the certainty of an immediate execution had diminished, we'd left the cave and marched fortified into the solemn night, swaying as we walked.

It was to take days, not hours, for us to accept the idea that we would live. It took even longer to justify in our own minds that we should survive. It's strange, but nevertheless an experiential fact that we ourselves didn't believe that we should've been allowed to live. It was obvious to us, as surely it must have been in all cases where men were captured, that large numbers of prisoners are disruptive even when utterly passive and submissive. Prisoners are in the way when it would be so easy to get them out of the way, as the German SS had concluded in the Belgian town of Malmedy, just a month earlier. It was that precedent we most feared but understood as as an element of total war.

So, we marched over the hills that night with stunned awareness that each step carried us farther from friends and closer to our foes. Those were weary, plodding hours and our reserve strength was gone. We began to grow a new skin to protect us from the new impressions, and like children we'd delighted in the fright of something that was altogether new. Soon, we would lose even that. The first light of day brought us to a stone barn close to the roadside, and we were herded into that building like so many willing sheep. Then came sleep and with it the release. When they awakened us thirty minutes later, we discovered that half of our number were missing, and not to return. Where they were taken, or why, we were not told. In the confusion of that moment, our tumbled minds informed us that the obvious was answer, and the specter of death began to haunt us once again. Those of us who remained had been spared, but for how long? Were we really fortunate to have been allowed to wait a while longer? What we didn't know was that those who'd been separated from us were in greater need of medical aid than those who'd been left behind in the barn. Nonetheless, our senses continued to repeat with certainty that, at some point, it would be death for all us prisoners.

We were given coarse bread; we were given water and we walked toward what to us could never be a destination. This was the first day after the night and no man could walk far. By midday we had arrived at a small camp, a compound of some twenty frame buildings encircled by heavy wire and guarded by men and dogs. This was a prison camp, temporary perhaps, not far from the lines of the fighting. We entered and were separated by rank. A fresh team of guards, all barking hostility and death, had packed us into buildings and ordered us to remain silent. There were ten of us in a room about ten feet square. Our furnishings consisted of filthy straw, and a bucket which was emptied every twenty-four hours. For light we had a single window. Later, even that was boarded up after a fellow prisoner on the outside tried to call in to us. That first day was given to deep sleep which carried into the night. Twelve hours after we'd been put into the room, the door opened only long enough to push in another prisoner before closing again. The new man joining our dark circle was a British flier who'd been shot down close to the camp. It was soon after his arrival that the interrogations began. We were called, singly, in pairs, in groups, once or twice or a dozen times.

We spent the rest of the night and most of the following day being unlocked from our room, locked into our room, or parading between the room and the building where many German officers sat and requested our confidence. This was where we prisoners were assigned numbers in lieu of our names. This was where our keepers inextricably decided to keep us on the side of the living, instead of the numbered dead. We were given bread, water, and incongruously, twenty minutes of calisthenics under the supervision of the Colonel Camp Director. Later that day we left the compound and headed deeper into Germany with increased numbers. We were now one hundred.

We didn't get far before being ordered to halt in an open field for the night. We did not know then, as the guards and dogs patrolled the perimeter of our field, that this was the beginning of a pattern that would be relentlessly repeated in the countless days and nights ahead. In the months ahead, our routine was to be one of perpetual walking by day. The nights were spent in damp, muddy fields where prisoners huddled together for animal

"...when U.S. artillery began raining down...the Germans delighted in the fact that Americans were being killed by their fellow Americans."

warmth and waited for the cold dawn of the new day. We were not properly outfitted for those cold, miserable winter and early spring months. In the very beginning our clothes had been taken from us and we were allowed to keep only one upper garment and one lower. It wasn't long before our boots wore thin on our wet, unsocked feet.

Thus equipped, we were to march a four-day schedule of twenty, fifteen, ten and five miles per day with the fifth day being a day of rest. Actually, the day never came when we had full rest, since we always had to march a part of that day also. We were required to do all of this on a ration of one sixth of a loaf of bread each day. Had this been known at the outset, it's likely that many of the prisoners would not have carried on in the way that they did. In a sense, we were saved by our imaginations telling us that the next turn in the road would install us into a camp with the semi-comfort that was supposed to be a prisoner's lot. It was never to be.

Four months were ahead, one hundred and twenty days of living with the hope and expectation that now, today, there would be rest, there would be food. Yet each day brought with it the disappointment of no rest, no food, and another rainy night in an open field. Nerves were taut and brought to the surface. The breaking point was always close. That was the basis of our life, such as it might be called, as prisoners of the Germans. Despite the threats and implications, this story does not begin to contain the truth of why a man can live, and what his passion is for his insistence to live. Ironically, it was the the persistent sameness of our days, the constant walking and lack of food, that had given us a sense that we may, after all, survive the war.

Naturally the season and the situation brought pneumonia into the lungs of all of us. I am sure we would have confounded the world's finest medical brains for breaking all the rules concerning the limitations of the human mechanism. Pneumonia came and it went; how many times it inflamed our bodies we could not say, only that some could not go on. For these, oxen drawn wagons followed our column of dispirited men and took the burden of those who could no longer walk. There also were times when even the wagons were more than a man could suffer. In those instances, the man would be taken to a hospital whenever we passed through a village or town.

We neither heard or cared about those who were left behind since we were too focused on the mystery and blight of our own suffering. The walking never ended; the torture could expand no more. Our numbers changed almost daily and in ways that could never be explained. At times we might be as few as fifty. In other weeks we might suddenly become five hundred or two thousand. What determined this, only our German captors knew. Two roads would meet, and they would split our numbers in half at the intersection. Where we had once been one hundred men, we now were fifty. Or, after joining up with another group, where we had once been fifty, we were now six hundred. At another intersection we would become three hundred, and so on. These transitions seemed to happen casually and made no sense whatsoever. Only one thing was always certain. It was forever the rule that we walked in groups of twenty-five with space between each group. One guarding soldier one chained, snarling dog were assigned to each group. This system of guarding us never varied, regardless of how small or large the number of prisoners was.

Of course, the guards did not walk in all ways that we did. They also changed constantly, the same men never being with us for more than two or three days. What determined this was also unknown, but likely our proximity to larger towns or army installations had much to do with it. Just as the lack of food and rest had introduced physical problems, the changing of the guards often brought on human brutality that we were in no condition to face. It's true that soldiers are stamped out in the same mold, but the factor of a soldier's personality and character is always a wild card. To us, this changing of the guard was just another threat to life added to our already heavy weights. No two men are the same or react to a situation in exactly the same way. So, it was the personalities of our guards that ultimately determined the treatment we received.

People forever ask, "how were you treated?" The answer lies in our everyday conduct, and in the contact we have with other human beings. One man will react to you one way, another another. This also was true of our captors. Every time we passed through villages where women would put out buckets of water for us to drink, one guard would say "YES," and another guard would say "NO." A third would just kick the bucket over with his booted foot. We never knew the guard who filled a cup for us and said, "DRINK!"

Part II
The Shutters of Consciousness

The days were all the same. An unbroken sequence of physical misery and mental depression. It did not happen at once; the process was slow. As I've said, we had expected to die and could not understand why those who had the power of life and death were dragging out the hours that remained to us. We knew that death was close, and we waited in despair for the inevitable. At some point in the days that followed, after we were given medical attention, interrogation and organization, it finally became apparent that we were not going to be killed. We had become accustomed to the sullen looks of some of our guards, to the humanity and inhumanity of some, and to many other things that happened in that first week. Then, at last it became apparent that we would not be chosen to die. It was at the instant of this recognition that we began to live the lives of gods. We would not be killed therefore we could not die.

When a man is given the faith that he will not die, he becomes invincible. He stands above those mere mortals of whom he was a member only a second ago. That it happened in our case was not as remarkable as it was inevitable. It was the only way we could have survived, the lone possibility. Our bodies were already feeling the strain being put upon them and anticipating the additional strains that lie ahead. It was this physical torture that helped our minds perform in ways we had never known. Independently of our own thought, the mind had made a compact with the body that if one would deceive us, then the other would too. By interacting with each other they would bring our dull senses through, even if they had to make us believe we were supermen in order to do it. It was perfect cooperation.

As events were to show, both the mind and the body were about to begin operating independently. To us, mind was to be utterly divorced from body, and body from mind. We couldn't hope to understand that a new plateau in physical/mental cooperation had been reached. Our conditional price of survival was that we never understand or explain what was happening to us. Our lives were to take on new dimensions and we were not to be allowed to know what they were. In time, the agony that we were to endure was to be on a level we would not understand. It was no sacrifice, not consciously at any rate, only that our sufferings took from us all hope of awareness and of heart.

If soldiers are unhappy, and who has known them to be otherwise if they have any intelligence, then soldiers who become prisoners are miserable. On the one hand they are successful in that they have cheated death, but the price for that is failure in the conditioned terms that the soldier has been indoctrinated to accept. To their friends who have not been taken, they have failed. This is a strange perception. There are some who will not agree, yet that same group would not logically follow through and say that success for the soldier is death after losing the friend who stands and fights beside him. By the time such thoughts come into the mind of the prisoner of war, he is no longer there. His mind has made the transition for him, and his body has descended to a level where his lack of alertness saves him from a degradation that none could face. Looked at from the physical side, the attitude of the prisoner is predictable. Certainly, there are enough medical records in this statistical world to show how the physical body reacts to the strain of starvation and malnutrition. These same records will show the limits that a slow deterioration can reach before death takes over and life ends. Inferred from the same process, the living tissue of the mind must also be fed or will undergo sympathetic changes and just as surely deteriorate.

Yet there is a difference, and it is all important. Which will go first we ask, the body or the mind? Or is death a simultaneous process that we have yet to recognize? Our meager knowledge tells us that life ceases with the body, not the mind. However, we have not yet found a way to ask the dead mind if the body is really there. We assume that the body is there and let it go at that. What if someone were to tell us that that body belongs to some other mind? I point to these possibilities only to give credence to what happened to us during the first weeks of our being prisoners. The limiting factor to any understanding of a group reaction was that you had to be part of the group. We'd undergone changes, both physical and mental, that no physician could understand or explain to us. The struggle to hold tenuously on to some reality, be it physical or mental, was personal. If you were unable get outside of your own conscious being, your starved awareness would eventually consume you. We were probably so concerned with impending death that we lost orientation with the small moments which are, after all, the meaningful ones. Having done that, we drifted comatose and lethargic through each day not caring what was was happening or what subtle changes were taking place within our own beings. When the division of mind and body had taken place, however, it was unmistakable, and terrible to recognize.

Coming from many places, having known a variety of environments, educated to different degrees, reaching out in diverse directions, it's no wonder that each man reacted differently to the forces that were working on us. There is no case to be made for the fortunate or the unfortunate. Those who could stand up better under the pressure of one situation failed to grasp the fundamental nature of an added strain. Situations that hammered with the force of an iron sledgehammer on one man would not be noticed by another. Conversely, some small stress that didn't touch the consciousness of many, would fall heavily on a few. Consequently, there was a definite consciousness of time that each man had before reaching the level where time, or anything else, mattered. We all all had similar periods of waiting for the pressures to gain momentum, bind us up into knots, and leave us to the mercy of our own mental punishment.

Even with the diversity to be found in the many prisoners, there was hardly a man who had not submitted to a captive's metamorphosis within four weeks. For some it came as early as two weeks and there were others who resisted for 5 weeks. Speaking generally, however, four weeks was the typical limit. Then, once the prisoner had stepped into the harlequinade, a boundary was crossed and there was no turning back. When the change occurred, the mind and body would fly in different directions, and the broken man would descend to an animal level.

Yes, "an animal level", and even as I say it, I realize the difficulty of the term. It means so little to those who have never crossed that line and it certainly does a grave injustice to the animals. Many use the term to describe humans who are below them on the social plane. With as much justification, those humans who are beneath the proud and the wealthy use it to describe that so called higher social strata. Here I mean only that I recognize both man and beasts as organized beings endowed with life, and that the functions of the beasts are presumably less refined than those of man.

Suffice to say that many of the men became as beasts, and in many cases were lower than the lowest of the animal kingdom. Considering the circumstances, I do not say that this was wrong or right, only that it happened and there were many reasons for it. Some of those reasons have been mentioned, such as the conditioning factors. Other reasons lay hidden deep within the individuals for reasons of fear, hate, or ignorance.

For prisoners who hadn't yet sunk to an animal level, the dis-

The Shutters of Consciousness

"...When the change occurred, the mind and body would fly in different directions and the broken man would descend to an animal level."

integration of another was a hard thing to witness. You refused to believe that this would eventually happen to you. Yet, you could not turn away from it no matter what your rules for self-deception might be. One by one, each man broke under the strain, and one by one each had a final shock of recognition before descending into the lower regions of his humanity. How well I remember the day when, as we walked down a road I examined my left hand. We had been prisoners for some weeks and by this time our routine of perpetual walking had been firmly established. One day was as the next with little variation, and our attentions wandered in order to deny the reality of the moment.

On that day, I don't know how long we had walked or where. I do remember being faced with the knowledge that soon I would join others and draw the shutters of my consciousness closed. I, too would set out on a personal journey by driving my thoughts into the subconscious. If this was courting death it was also the means for sustaining life. It was the instinct for survival and the best way that any of us had for ensuring our return home to the States. My left hand, as with other parts of me, was becoming less and less a thing that was mine. It was an attached something that went wherever I did for reasons that I no longer cared to explain or seek explanation for. But on that day, it asserted itself and demanded recognition from me.

As we walked down the road, I studied that hand and found that my fingers could scrape furrows in the dirt that encrusted

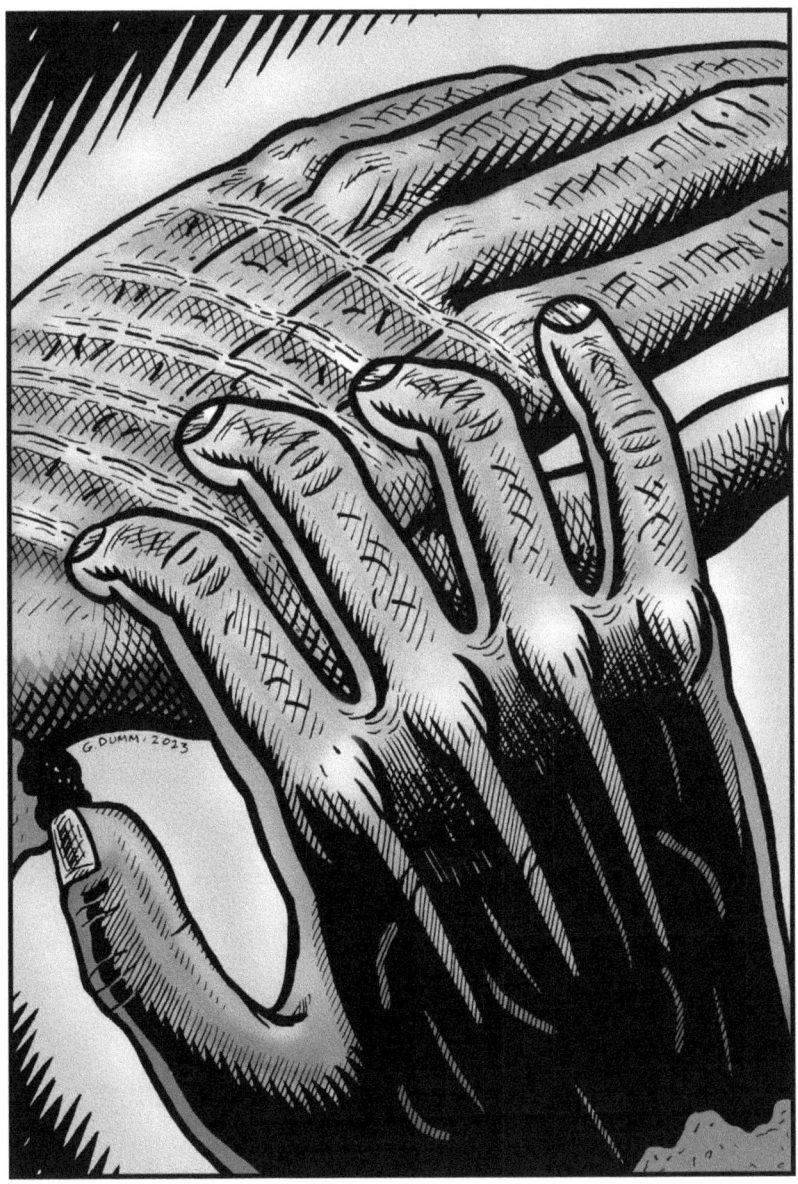

"... I was held in fascination by the fingers that could scrape ruts into that dirt, and make it look like a ploughed field or ...a chain of foothills."

the back of it. From a great height I looked down on it as you might look down from an airplane on some brown, contoured, ploughed field. I knew that this was my left hand, and I was overwhelmed by the hopelessness of all the dirt that covered it. Still, I was held in fascination by the fingers that could scrape ruts into that dirt, and make it look like a ploughed field or a jagged chain of foothills and mountains rising above shadowed valleys. Seemingly, for hours I marveled at the sculpted topography on the back of my left hand before realizing that it was the fingers of my right hand that were creating the effect. In other words, my two hands were becoming less and less things that were part of me. And in that final moment of recognition, in that spinning instant of time when I realized that my hands were still mine, I knew that inevitably the blinds must come down, and the doors of my consciousness must close.

This was only one of many perceptions, and each was as ugly as it was fascinating. I wasn't the only one experiencing such delusions. They were happening to the others as well. None of us understood precisely what it meant to anyone else other than himself. In the beginning you wondered about it, but never spoke of it. For days and weeks, you knew it was all around and only a matter of time before another man succumbed to it. Knowing that sooner or later it would come for you added weight to each man's misery. Finally, when the day came that your preoccupation was reduced, and the angle of vision was brought into where you could only sense yourself, the process was finally complete. Though you walked in the company of hundreds or thousands it did not matter any longer. Now you were alone, the last alone man in all the aloneness that those hours could contain.

Part III
An Affirmation of Spirit

We forced our bodies over the roads of Germany day after day while following routes that lead nowhere. Often, we would cover the same ground or arrive at a recognizable point two or three times. Ordinarily there might have been some concern or curiosity about why this was so, but in the flattened minds of the prisoners it really didn't matter where we were or were going. What was happening was very simple. The German army, being pressed hard by the British, French and American forces in the last few months of the war, was feeling the confusion of disintegration. From all sides the Allies were moving in, sometimes at an incredible rate of speed. Cities were being overrun. Army and prison camps had to be evacuated almost overnight. Transportation had been brought to a standstill because of an aggressive Air Force overhead. Confusion reigned, and it was difficult to say which was the way to turn. For us constant walking was our lot, not only because of POW camps being moved, but also because by that time in early 1945, the prison camps were overflowing with Allied soldiers.

Every man walked alone and was concerned only with himself. Brought to the breaking point by exhaustion, the prisoner found that his movements were mechanical and were motivated most often by the commands of a guard who spoke in an unknown tongue. The prisoner cared nothing for the man beside him. He did not speak, and his eyes seldom left that spot on the road where his weary foot would fall next. The world had shrunken as small as it could until its outer boundary was the limit of the individual's skin. For most of us, the mind was on a holiday and only physical sensation could awaken us to the fact that our lives were hanging on by a thread.

The greatest concern for all was food, and if anything could remind a man that others were suffering around him, it was the daily bread ration. That was the only time there was any spark of life in the prisoners. That's when the prisoners came out of their dark shells and stood up once more, not as so many floundering pounds of flesh and muscle, but as individual humans. True, it was not occurring as an affirmation of man's goodness; nevertheless, I felt something sustaining when that mass of men would come alive and be counted as individuals.

There was nothing for anyone to be proud of in the actions of these men. I wondered if the sons, husbands, or lovers who knew these men would recognize the depths to which they had fallen? Would they deny that such a thing could happen to others? Perhaps, but not to those who were close and had shared the living moments of their life with them. Each man had been graced by the covering cloth of civilization. His sins, if such they were, received absolution in the deluded heart of someone who placed him above his fellow men.

Who were those men who received their sixth of a loaf and stuffed it quickly inside their grimy shirts? The same men who separated themselves from the herd and ran to an isolated corner of the universe. The men who growled, saliva running down their chins as they tore at their meager ration with gnashing teeth. Who had held these men close at night and loved them? What mother had seen this soldier growing straight and tall toward honest manhood? Who were the men who took their bread, broke it into three pieces so that it might represent three meals and while quickly eating one piece stuffed the other two into the crotch or in the pit of the arm so that it wouldn't be stolen?

What was their preparation for this? From the north and the south and the east and the west who were these men? What was the life which had taught this one to push his mouth so full of bread that he couldn't possibly chew it and would have to spit the whole half-wet mess out again in order to start again? I clearly remember the sadistic prisoner who silently ate his bread crumb by crumb, reverently, slowly, silently. It took him hours to do it, and all the time his eyes were watchful, and warned the others not to come close. Oh, he was not so innocent this man who had once driven a truck, or made steel, or counted money in a bank. He had waited and watched while the others had eaten. Now was his moment of meaning: let the bastards watch, let them drool, let them try to come close. Crumb by delicate crumb he ate, and his nourishment never came from the bread.

What a man did with his bread was his own concern. How he approached the daily ritual certainly mattered because what

An Affirmation of Spirit

"*The men..., saliva running down their chins as they tore at their meager ration with gnashing teeth. Who had held these men close at night...?*"

one of us did, we all did. It was something we had to think about to whatever degree we were able to think. Methods for eating a portion of bread may seem of small importance. To the prisoner they were not. It was the supreme moment of each day, and the way in which bread was consumed affected all of us, including the German guards who watched. It was just another of the many chaotic situations our hosts created so that every miserable one of us could find reasons to turn on each other.

Somewhere beyond the Rhine we came to a very large prison. In it were housed more than four thousand French soldiers who had been captured in the early part of the war. One wing of the building held two thousand British prisoners taken during the days of fighting in North Africa. For us who arrived that day there was the hope we would stay for a time. Not that prison is very attractive, but for us it spoke of rest, and, perhaps, more food. As it happened, we were there for two days, the longest we'd stayed in any one place. For quarters we were taken to a stone building separated from all the others by barbed wire fencing. Two hundred of us were herded inside where we found long wooden platforms which had been laid over the top of forty or fifty horse stalls. The floor was brick, covered with straw with an accumulation of horse dung. Needless to say, the building was rotten, smelly and infested with vermin. However, to us it was a roof and more comfort than we had known for some weeks. This prison remains with me because of three remarkable things that happened in the short time that

An Affirmation of Spirit

we were there. Two were humorous, a third was a positive manifestation of life that touches every man, even if a good number of them cannot fathom such mysteries, because they have never been able to tolerate the mystery of themselves.

The first humorous incident occurred when I tried to barter for a toothbrush. Speaking to a Frenchman through one of the fences separating us I learned that he could get me a toothbrush. It was his idea, not mine, for though I had then been without a toothbrush for some weeks and would certainly find one more useful than the frayed ends of sticks that I had been using. I did not honestly think it was an item that would take precedence over many other things I might need. Yet, he was not offering food or soap or any one of many things. He was offering me a toothbrush. It was a welcome suggestion, and I tore another strip of leather from what had once been the top of my boots. It is strange to think that a toothbrush at that time suggested the greatest of luxury. Weeks without one, for those accustomed to their daily use, leaves the teeth in a condition that must be somewhat similar to the working parts of an internal combustion engine. Now that I was about to have a toothbrush to use, I became, in my own eyes, the most fortunate of men.

I suppose as I waited for my friend the Frenchman to return that I was planning the campaign I would follow once the toothbrush was in my hands. Such things as when I would use it, how often, and whether I would hide it from my fellow prisoners or share its comfort with them. For some time, I waited by the fence trying to look innocent in the eyes of the guards for we had been forbidden any contact with other prisoners. At last, my friend returned and dropped his prize through the fence some short distance away from where I waited. Trying not to show haste or interest in anything close to the fence, I warily approached the spot and secured the prize. At first, I didn't dare to look at it so I slipped it in a pocket and went inside our building to study it in private. At last, it was in my hand, and I looked at what was undoubtedly the finest toothbrush ever made by man.

But was it? I held in my hand a black handled object that had a tuft of matted black hairs projecting from its end. These hairs were not many in number, perhaps one-tenth of the original number, and they were so caked with grease that no amount of separating with the fingers would allow a single strand to be moved away from the mass. Ugly and uneven, no two hairs were the same length, and all were embedded at the base in about a quarter inch of black dirt that had hardened like concrete. This, then, was the prize, The tool that was to clean my teeth and put me back into the ranks of the civilized. Yes, I kept the toothbrush for time but never used it. One night as I slept, one of my good American comrades stole it from my pocket, and it vanished forever.

The second humorous incident at the prison occurred during the night. It was habit for the Germans to lock the heavy barn doors of our quarters about five o'clock in the evening and to leave only one guard outside. He was posted by the door for the purpose of letting us out when we had to make a call to the toilet which was out in the yard behind the barn. This toilet house needs some description. To begin with, it was used by anywhere from one hundred to one thousand men. It had been built from rough-hewn logs. It was a single unit some sixty feet long by perhaps thirty feet wide. Upon entering from either end, there were only two large doors, you found yourself looking at a gigantic concrete trough which was only slightly smaller than the full dimension of the building. This trough was eight feet deep and was flushed twice a week by what looked like a ten-inch water outlet at one end, at the opposite end was an escape door about four by ten feet. Around the perimeter of this concrete basin was a log fence about about knee high. This was the toilet we used.

At night of course there was no light in the yard or the building. Sometime after midnight I pounded on the door which the guard then opened with a grunt. I took myself outside to the toilet house. I had no more than entered when I heard a thrashing down in the concrete trough and the voice of a man who sounded farther off than he really was. "Give me a hand-out - give me a hand-out." he kept saying, and then because I hesitated from surprise at the uniqueness of this situation, he added "Or are you another of them bastards who won't give a hand? I been down here for darn near an hour," he pleaded. Of course, I helped him out.

Both the toothbrush and the man in the toilet were what I would describe as humorous incidents. What happened the following night was to have dimension to it and provide some of us with enough energy to carry on for a long time. It's a fact that the prisoner's life is a series of blunt degradations. If, somewhere along the line an occurrence brings sanity and dignity back into focus, then a man can carry on for quite a while. He is sustained and nourished by something of equal importance with the food he eats and the water he drinks. The hour doesn't matter, but say it was ten at night when the guards opened the door to our large barren cold room. It was the second night and many of the men had taken to sleeping on the floor after their personal experiences with lice the night before up on the boards. It was no better, but there was the illusion that the filthy floor of a barn might be. However, lice are lice at any level. You could take your choice and be eaten on the boards or on the bricks.

In the darkness you couldn't tell what was going on when the door opened but someone heard the guards saying that they were putting four new American prisoners in the room. You would think there would be some interest in new men, but there wasn't. Nobody cared who came or went. It would have meant nothing if the guards had said they were putting the whole Brooklyn Dodgers baseball team in the room or the then Secretary of the Navy, James Forrestal. No man could sleep, and each was in his own cocoon of fog. The four new men fumbled in the darkness looking for some place to try and sleep. Muffled curses, kicks and groans filled the air as they moved about disturbing men. Then they settled in a corner.

It is difficult to describe the quiet pandemonium in that horse barn. Discomfort would have been one thing, discomfort in the company of several million starving fleas was quite another. To men who were without patience or tolerance of any kind it was a tossing, hellish night that had to be traversed with curses. So many men were caged therein that the effect was like the human equivalent to a turbulent river over rocks inside a deep canyon. Anyone could have tried to relieve the situation if they had known how or had the energy. It fell to the lot of the four new men to try it, perhaps because they had more energy or hope than the rest of us.

It started as only a rhythmic whisper in the high air above our heads where we could not touch it. It rose from that slight murmuring to a soft, delicate humming, always in harmony. Soon it picked up, the rhythm became stronger, the harmony more accented. The soft strumming began to fill the fetid air and while it was persistent, it was never insistent. Yet, I suppose it was bound to happen that the men got restless, more out of fear than curiosity. This was intrusion, it was the beginning of a demand for participation that no men wanted to answer, not sympathetically at any rate. So, they began to complain in the darkness. By this time all four voices were individually clear and were singing a melody of quality and beauty.

The singing men rolled along in wonderful song, but it was more than the men could take. They cursed and shouted and soon the guards began shouting to stop the shouting. Silence invaded the room, and the air which had begun to feel the purification of song became rank once more. Once more the harmonizing voices tried to find acceptance. This time it was a religious song that was more imaginative than religious. The room began to fill with four voices full of song. Now the voices were sure of themselves. It was as though they knew that the initial protests had been an almost involuntary reaction from strained nervous systems.

Now there was quiet as the four voices in harmony moved into song. Here and there in the darkness a single protesting voice would tell them to stop, but for every voice that said stop there was another from some other dark region to defend the singers with an affirmation, a protest against the protest. Soon this byplay involved those who cursed and shouted, "shut up you goddam niggers," and those who balanced the night with, "quiet yourself, and let them sing." It went on that way for about two hours; angry voices calling back and forth in the darkness and the quartet of singers moving from one fine song to another without a pause or repetition. Until the small hours of the night the strength of song continued, forever being balanced or unbalanced by the strident voices of the men in the room who tried to tear the foundations from under it or to hold it up for what it was. That night those songs coming out so easily and so softly, as I have said before, restored some dignity to the mad world in which we moved. Yet even that precious gift was unaccept-

An Affirmation of Spirit

"...these men who walked close to the edge of losing faith, who hungered for food...who knew only the pull of the outgoing tide, began to sing."

able to many because they were conditioned to discriminate against the skin color of some of their fellow soldiers. It was not the song which infuriated them as much as it was the men who sang the songs to soften the bitter hours. They could not accept what was superior to anything they could have given themselves because, in their own warped minds, the givers were inferior to them. This was the blackest of insults; answer it only with the curse and shout.

With the first light of morning, I think every man was alert to what had happened during the night. There were those who wanted to see the men who had been singing, perhaps in the recognition of the eyes to thank them. And there were those who were alert because they anticipated or wanted to finish in the day what had been started in the night. Actually, it was to turn out quite differently, for the singers of the night anticipated that there would be this reaction and they had prepared themselves for it. They wanted no trouble. One of the four came to the middle of the room almost as soon as the first men were stirring. He was a tall man, smiling and clear eyed.

He spoke and his voice was recognizable from the songs of the night before. "I just want to say a few words, " he started. Immediately he was shouted down as a "nigger bum" and a few other things. Almost at the same time he was defended by those who had held the protesters at bay in the night. Five times without moving from the spot where he stood, he repeated, "I just want to say a few words." While the room seethed with wild and distorted argument, he waited for the majority to win before speaking. I could not, in justice, repeat the few and simple words that the man used. But his thought was straight enough to penetrate every man in that room and to turn those who needed it back into human beings. He told the story of himself and his three friends, and how they became prisoners. He told how we all shared the same fate and had to deal with the situation we were all in. He mentioned that in the United States negroes had not been allowed to participate in the white American society that they were now being called to do. Yet there was no added bitterness, no added strain, they were proud to be among us since fate had made it so. He spoke of an equality that was fact and could not be altered by social prejudice. If we were to fight side by side, be prisoners side by side, perhaps die side by side, why could we not live side by side?

That was the question that he posed some hours earlier, in the songs of the night. How did the group react to what he had said, for react they must have? Except for the sullen, the weak, the lost, they did react. They all cheered the man! Before we all turned back into being prisoners once again, these men who walked close to the edge of losing faith, who hungered for food and more than food, who knew only the pull of the outgoing tide, began to sing.

No individual started it, no man thinking back on it could say that it was in his mind. It was spontaneous, as natural and wholehearted as anything that men had ever done. It was an affirmation of spirit, the recognition of life as being in each and all of us, and in that instance, our quartet of negro singers were in the center of it carrying the rhythm that no man could deny. How long it continued, for minutes or hours, is a mystery.

It ended when German soldiers burst into the room, cursing the American pigs, and shoved us all out into the cold reality of another long day.

Part IV
The Stench of Fear

The weeks wear on and the prisoner of war becomes more and more removed from himself, while in his few lucid or intimate moments, he becomes closer to his essential self than he has ever been in his life. Few circumstances lift him above the misery that weighs him down, but it is those same few circumstances that are wonderful for they bring release and allow him to walk closer to the ground where other men walk. Perspective becomes meaningful and purpose is remembered. Suddenly, the lake where he is slowly drowning does not seem so deep.

For what it's worth, the secret of life for the prisoner is having the energy to be aware of human movements and motivations taking place around him. Such awareness can keep away the hopelessness that circled each man like a vulture in the desert sky. Yet the energy to perceive did not always happen at the same time of an event's happening, and it was difficult to stay aware of the vital things happening around us. I have two examples that will illustrate the nature of these opposite necessities. In neither of them do I intend to make commentary on the moral nature of the deed described.

The sky over the German town was oppressive with bombing planes. The people had safely retreated to the underground shelters while their homes rocked wildly above them. When we walked into the town, our entourage consisted of three hundred prisoners and eighteen guards. The bombing was just getting underway, and our guards, being reasonable men where their own safety was threatened, stopped our column at the first shelter. It was a tunnel cut into a high hillside on the outskirts of the town.

Undoubtedly the tunnel must have gone far back into the earth. It held many people, and it was already close to being full. The guards stopped and asked for shelter in such a way that they included the prisoners who were with them. Some minor official of the town was at the entrance to the shelter, and he received their request as though they were madmen. More than that it made him downright angry. His face turned red, his nostrils twisted, his eyes bulged, his chest heaved. He hurled wicked sentiments at us through the noise of the British air raid with all the force of the bombs falling on his town. There was nothing our guards could do. They were frightened, but had to lead us on. Had the official been a different man, the shelter larger or less crowded we would probably have stayed there and the event that I want to describe would never have occurred. Such were the movements of those days, though again I will say that to the prisoner of war they fell in no such sequence and could not chain themselves to any meaning.

From the streets where bombs fell, we were taken to the basement of a school that must have been a university, judging from the number and variety of buildings. We were locked in what was the service part of a good-sized installation and all around us were the massive pipes which carried hot air from a furnace to the many buildings. I do not remember that any of us were impressed with the heat of that basement, and that leads me to believe that the furnaces were cold. Weeks we had been cold and wet, if suddenly we were shut up in a warm furnace room there would have been those effects to remember. None remain, so the basement must have been cold.

Within the hour the air raid ceased, and it was perhaps another hour before guards came to our doors. They explained that we would be taken out in groups of twenty-five men, but they did not say why. Much was my own surprise when I left with the eighth or ninth group and found that we were indeed on the grounds of a large school. Once we were all assembled, they walked us across a quadrangle and into a dining hall. To most of the prisoners this was more of a shock than if we had been walked into a lime pit.

Upon entering the hall we hesitated like wild animals would when brought to the door of a cage. Our guards kept us moving toward a serving counter and it was only when a white china soup plate and a spoon were in my hands that I felt an extreme embarrassment at my dirty and lousy condition. Somehow there was profanity in taking these filthy men before clean counters and steaming food. It was cruel to the men, though our good intentioned guards were certainly not guilty.

Before us was food in a variety that we could not possibly comprehend, and none of us made a move. None wanted to be the man to take the hurdle which would land us all back in the cold, wet, and hungry world from which we had come. Stupidly we stood before the steam tables in silence and waited for the vision to pass. We were sure that once we had seen the food, our guards would tell us it was all a joke, and now we could get the hell out of there and back on the road. To break the spell a guard gave the first man a push forward and one of the cooks held up a ladle of hot potato soup which found its way into his bowl. We all followed as though hypnotized. One small ladle of hot potato soup for each man. It mattered not that there were also sausages and hams and great cauldrons of noodles in the steaming pans. For us, it was soup, hot and good, which we could not believe we were eating, even as we tasted it.

Across the room from where we ate four German officers were finishing their lunch. They were relaxed and occupied with a conversation that obviously was all about the prisoners. They kept studying us, nodding our way and generally making us feel miserable because of our rotten condition which contrasted so mightily with the hygienic dining hall. So, we hurried our soup with the thought that we didn't belong in this place and would do best to get out as soon as we could.

Then, a strange thing occurred. I had finished and was waiting to leave. Our guard came over and ordered me to have more soup. He didn't even wait for an answer but pulled me to my feet and gave me a shove in the direction of the counters. As this was happening, I noticed that several of the others who had already

The Stench of Fear

"One small ladle of hot potato soup for each man...For us, it was soup, ...which we could not believe we were eating, even as we tasted it."

finished made a move to get up and go with me for soup. This was not to be. The guard pushed them down as forcefully as he had dragged me up. What it meant I could not know. I went to the soup counter and held out my bowl before one of the cooks. I waited for the ladle full of hot soup to rise into the air, but no ladle came, no soup came, and in its place came the wrath of the cook.

He informed me that standing before him was one of the most ungrateful of all the ungrateful American pigs who ever soiled the earth of Germany with their foul boots. He spit to the floor in front of me and grabbed my bowl. He would have broken it over my head if suddenly there had not been a shout from one of the officers sitting at the table across the room. This officer rose from his chair and stalked militarily erect to the counter. He picked up the cook's ladle, took my bowl from the shaking hand of the cook, and proceeded to give me a ladle of soup. He nodded, turned away and walked back to the table where his fellow officers sat watching him.

For myself, I was too confused to do more than stumble back to the table where the prisoners were seated. On the way back to my seat, I looked toward the table of German officers. I was hoping that my eyes would meet those of the officer who gave me soup, but he was in conversation with his comrades and his composed face looked as though nothing had happened. The same was true of our guard who had pushed me forward. He also looked impassive, as though what had just occurred was nothing remarkable. In my case, I ate the soup without any enthusiasm and was unable to meet the eyes of my fellow prisoners with whom I had been unwittingly used as an actor in some weird stage play, the nature of which I knew nothing.

It was afternoon by the time the prisoners were finally re-assembled in the town square. We had felt a reluctance in our guards to take us out into the open countryside once again. It was a town of good size and the guards naturally wanted to stay overnight. However, to do this they would have to find some space where the prisoners could be contained for the night. They appeared to be planning this while we waited along the town's curbstones. At last, some of the guards who had gone off, returned and apparently the matter was settled. They assembled us into groups. Any group of prisoners marching through a German town aroused curiosity. People came to the sidewalks as though for a parade, and they brought an air of festivity with them. In their emotions we evoked the full scale of reactions, everything from killers to unfortunates. In fact, the people of Germany had reminded us on countless occasions exactly what our presence meant to them. It was hard for us prisoners to see what interest could be contained in our bedraggled conditions, and we could not recognize why these people viewed us as they did.

We marched through the town and then to an enormous railroad yard which, judging by its size, must have been of great military importance for Germany. We were taken past a large station house and loaded into freight cars that had been lined up on the sidings. Once sixty prisoners had been placed into a car, the doors would clang shut, bolts would slip into place, and heavy wires would secure the doors. If nothing else could be said of those cars at least they had walls and a roof, and this was protection from the wind and rain.

There was no comfort inside the car, however. Sixty men jammed into a freight car are well crushed for space. There is not room for all to lie down. If one man does lie down, another must stand. At best, we could all sit if every man was careful to bring his knees close to his chin. Naturally, there was not that kind of order or cooperation. As usual, it was every man for himself and that meant a situation which was destined to be bad. Nor was there any relief to contemplate.

It was late in the afternoon when they locked us in. This meant the boxcar doors wouldn't slide back until the following morning. Description will not bear witness to how uncomfortable men can be under such circumstances, or to what extent they themselves add to the burden by being unreasonable. It was the selfish thinking only of oneself that contributed more to the misery of those stifling hours. Up to this time of our captivity, with one or two exceptions, the prisoners at least had room in which to move aside so as not to identify with the men around them. Now there was no possibility of such escape. The breathing of one man affected the lungs of the man beside him. The slightest quiver of a nerve or movement of a foot could immediately force three others to react. Before the black of that night was to put down the day, these occupied boxcars had already reached the explosion point. One more incident, such as a hand being stepped upon, would light someone's fuse and mayhem would ensue.

We would be saved from such self-destruction by two things. The first being utter exhaustion, and the second being the dreaded return of British Lancaster Bombers on a night mission. Through their wasteful struggles for space, the prisoners had brought themselves to that bankrupt air where all energy has been expended. It was hardly more than dark when the freight cars finally fell silent. There was no more to be said, no more energy to lick and curse with. The legs simply had gone out from under, and the depleted men fell to the floorboards in heaps of frayed nerves. There we waited for the hours to pass, tossing and moaning, finally accepting the crushing against one another as a purely impersonal thing. Hours under such circumstances are long, terribly long.

Looking down from their height on that row of loaded freight cars, the stars of the night must have mistaken the rising moans of men for the saddened murmurings of lost animals. A miserable menagerie collected from the far stretched corners of the earth, inside we had no water, and our toilet was the scant crack between the doors and the bed of the car. What light there was came through chinks in the roof and the quarter inch separations between the boards that formed the walls of our prison. Though the night light coming into our car restricted sight, it was kinder than the light of day had been. Yet suddenly there was much too much light. Unrelated to the hour, it was unnatural and hurt the eyes with its synthetic glare.

The hour was 11pm and sixty bodies were alerted as though they were strung to a single mind. Overhead the light continued, and as we heard the steady, pulsing drone that announced death from the sky, we knew the light must slowly float to earth. Soon it was joined by other falling lights, white revealing, terrorizing with their silence. But behind these rings of quiet light there was the steady insistence of the men in flight, the wing borne creatures who were making calculated adjustments, calmly preparing for that moment of glory when the bomb carrying warbird would sense its moment of reality, and the bombs would fall away toward the earth. Thus, in our freight cars, held to the fragile earth, caught and bound, helpless and hopeless, sixty pairs of eyes sought for the light revealing cracks that told us what we did not want to know. Flares floated above the railroad yards, they spiraled and turned with all the fun of a carnival; and they descended around us, before us, behind us and on us. Above, the British planes broke their tight formations and running down a celestial azimuth prepared to drop their bombs.

With the first realization that pathfinder bombers were overhead and dropping flares, the prisoners began to scream in terror. I

The Stench of Fear

"*Thus, in our freight cars,...sixty pairs of eyes sought for the light-revealing cracks that told us what we did not want to know.*"

say scream because that was the initial noise they made. It's fair to say that every man made some kind of noise that was beyond the comprehension of the man making it. How many screamed is irrelevant, but it was a long, prolonged scream by men who knew from experience that nothing could prevent the night flares from marking their targets. Picture sixty dulled minds groping for the insane limits of this incident in the flare lighted darkness. Senses were blunted, else they will feel more than just the agony of fear. They will feel the knives that were cutting into their senses, exposing fear, and giving vent to the human expression which has immobilized men when such things have taken place. I do not say coldly that men should annotate the dimensions of emotion when they are consumed with agony. I do say there is something better than what happened to these prisoners of whom I was one in every sense. I readily admit that none of us were able to give that something better, and that is the shame of human degradation brought to all human beings by the debasement of total war. If the stench of fear can turn men into monsters perhaps it can also deflect falling bombs. The shrill cry from the sky announcing the first bomb reduced our freight car of prisoners into a terror-stricken mob.

The first moment was suspension of sound, not silence, in preparation for the assault. Then, like a tidal wave smashing against some tropical island the men broke loose. They battered the sides of the boxcar, they clawed at the boards of the floor. They tried to hide in corners of the roof. They whined like broken backed dogs and slithered across the floor like snakes pursued by the mongoose. They tore at their fellow prisoners, screamed rage at their own feet, somersaulted in the air or sat stone cold on the floor. Some prayed for a long-lost youth, others aged so terribly that their fathers would have been as sons to them. Some sobbed and bit hard on their lips until the blood flowed down their trembling chins. Hardest to bear were those who laughed the high-pitched maniacal laughter of those who have passed beyond fear and danced like harlequins about the floor. As with other things this could not be sustained for long.

Finally, we paid our ultimate tribute to the terror with mute fear. The only danger remaining was the vibration in the ears for that alone could tell a man whether the bombs were coming close or falling further off. By 2am, the bombers had finished and left us not as sixty men, but as so many pounds of flesh and blood. Aching muscles and scarred tissues were held together by the glues of bodies wastes diluted in pools of fear's sweat. At dawn, when the Allied fliers came back in fast attack planes to survey the smoking ruins and machine gun anything that moved, we prisoners were too exhausted to react. Sky and the new day belonged to the warriors of the air.

After they'd shot up our boxcars and killed six men, there was no perceptible reaction. Men simply moved aside from those who died. The planes passed out of the air, the sun moved higher, men began to stir, and the stricken city began to move. At 8am the guards unlocked the cars and we tumbled out for bread, for water and to walk onward, eastward, into Germany.

Part V
The Inhalations of Air

A soldier is instructed that if he becomes a prisoner of war, it is his duty to escape, insofar as it is possible. There have been some fine escape stories told and they always demonstrated the determination and courage of the men involved. There is something of the foolhardy in all of them, and it is good to remember that they are the one in ten thousand. After we had traveled many miles deep in Germany, few had any heart for escape because what little heart they had left was preoccupied with survival.

The only serious thoughts that were ever directed toward escape in our group was when we approached within a few miles of Lake Constance in southern Germany where Baden, Wurttemberg and Bavaria meet. Then, for anyone with sketchy knowledge of geography, escape might have been possible since Switzerland, the neutral land where many doomed aviators were interned, was on the lake's southern shore. A few of our number thought that if they could slip away and find a means to get out on the lake, it would be possible. None had tried, however, for none were sure enough of where the lake was, and none had the energy to do it even if all other conditions were favorable. In only two cases did I know of men who tried to get away from the German captors and both of them were futile attempts. One was an idiotic demonstration proving that some men were more thoroughly indoctrinated as soldiers, sacrificing their own good sense to assume a station and authority they had never known in civilian life. The other case was more amusing than anything else and probably never meant to be a serious attempt at escape. It was just an outing, a few precious hours off from being a nonentity.

In our ranks was an aggressive United States Army Major. He was all bluster, something along the lines of the officer that all men hoped would fall overboard when troops were being shipped overseas. As a prisoner, he adjusted in a manner akin to a frog in a hot oven. To him our guards were stupidity incarnate, weak men who didn't dare to give him enough food to eat for fear he would break them apart with his bare hands. His actions were something like a cheerleader in a football stadium, except in this case the stadium was empty of people. Still, he carried on in ways he thought his high rank demanded. He was sure that he would escape and get back into the fight, even if it took him weeks or months to do it. His first opportunity came one night when we were all pretty fresh; the relentless, monotonous days hadn't dragged us too far under.

Usually, we never marched at night for the good reason that it made guarding us difficult. However, there were times when it was necessary, since there was no suitable field close by in which to leave us. This night it must have been two hours after dark when the major made his break. Each hour we stopped for a 5-minute rest. When that period was over the guard counted his twenty-five men, gave a whistle signal to other guards along the line, and we walked on. It is easy to see that a man attempting to get away would have five, or less, minutes of freedom before his absence was detected. Such was the case with the major. We had no more than fallen to the ground for our five minutes when he crawled off into the brush. When our rest period ended, we were counted, and the major was missed. Then, we were counted a second time. When they determined that the man was missing, a demonstration of German military efficiency was enacted.

Our guard gave a signal on his whistle and immediately half of the guards came to his signal. The others now spaced themselves to guard fifty men. Another whistle signal and each of the guards released the leash that held his dog. Soon a pack of Dobermans and German shepherds came trotting up to the whistle. From then on it was easy. The dogs found the trail, the guards followed the dogs and off they went into the night. The first time this happened they brought the major back in less than 20 minutes. From then on, their record improved since the dogs got to know the major's smell and the guards became familiar with the way the major thought and moved. The first attempt brought only reprimand, the second a good cuffing, the third a mild beating, the fourth a severe beating. On the fifth and last occasion they carried him back and presented him to the prisoners so that they might carry him until he recovered. The major was stubborn, and as consequences had proved, stupid. I remember he always boasted that he would never stop trying, that he would take the opportunity when it presented even though they might be marching him down the Unter den Linden. The beating he sustained after his fifth attempt to escape had finally convinced him to stop trying, despite subsequent, and perhaps even better opportunities to escape.

The prisoner who did get away for a time saw no reason to say good bye or why he was going. Even the prisoners who were closest to him never saw him leave. We had been marched into a town early in the morning, and our guards stuck us inside the walled yard of a bombed-out apartment building. It was a good enclosure for guarding since a high double iron gate was the only way out. Two Germans could easily control four hundred prisoners by standing just outside the building that was surrounded by a smooth eighteen-foot wall that no man was agile enough to climb, or so our captors thought. This was their conventional wisdom as we sat waiting in the rubble-strewn yard, as usual, for nothing. Meanwhile, the other guards went off to enjoy the town. They did not return until about three in the afternoon which left them enough time before nightfall to march us out of the town and into the open countryside.

Following habit, we were counted before the departure. In this case we were not in groups and counting four hundred men takes some time. On the first count it seemed only in error that there were 399. The second count did not round off the figure. When the third and fourth counts were made the guards looked at the high walls, the iron gate, and counted again. After six countings they closed the gate again and began a search through the town. Since it was not a village but a town of say seven or eight thousand people,

The Inhalations of Air

"...Two German officers had stopped the escaped prisoner beside the road and after a round of questioning they offered him a lift..."

they did not have an easy time of it. In the late afternoon it began to rain, and by 10pm we were in our familiar state of wet misery.

At 11pm, our gates were forcefully flung wide open, and amidst great shoving and shouting, one man was propelled by eight or ten boots back into our compound. The missing prisoner had been found. His story was simple. He had climbed the wall and wandered into the town looking for food. Luckily, he spoke fluent German and was hardly recognizable as an American. Though he was wearing the uniform of an American soldier, the stains and tears of travel had given him disguise.

He approached a house on a quiet street and identified himself as soldier of the Wehrmacht. He was fed a good meal by a solicitous German householder who'd sympathized with his having strayed from and lost his military unit. Given fat sausage sandwiches to travel with he set out. Our man showed no lack of ingenuity, he was having fun for he had no real hope of travelling back through several hundred miles of Germany to reach France. He asked for no quarter but chose a route out of town that carried him across a bridge guarded by soldiers. To them, he gave proper salute on either side of the bridge and left the town.

He became tired of walking and decided to get a lift from one of the many military vehicles leading away from the town in the direction of France. Two German officers had stopped the escaped prisoner beside the road, and after a round of questioning they offered him a lift toward the German border town he had mentioned. I think he went some twenty miles with the two officers before they came to the intersection where they left him. He was now in the countryside, in the free and open air, and again he started walking. Before long he passed a farmer working in his fields and this man shouted a greeting and asked where he was bound. The prisoner gave answer and continued down the road but apparently the farmer was wiser than the German Army. This was borne out when the prisoner came around a curve in the road about one-half mile from where he had seen the farmer and met this same farmer again. This time, however, the farmer had brought two friends with him. They were breathless and anything but friendly. Their first menacing request was for identification papers, and since there were no papers to show what was our man to do? These farmers came armed with one shotgun and two pitchforks. The escaped prisoner did what amused him by speaking English, even though he knew it would be his undoing.

This ended the prisoner's holiday, and the events that followed were routine. He was taken to the town jail for questioning, then he was sent back to be rejoined with his fellow prisoners and the prisoners' misery. He had enjoyed his hours, his food and the inhalations of air that tasted of his freedom, however brief. His only regret was for the one bad mistake he had made. When the soldiers came to apprehend him at the farm, they took the two fat sausage sandwiches sitting in his pocket. This small loss was a token price to pay for the hours that he'd stolen. For the brief time he'd walked alone, at least he was alone, as opposed to being under the constant scrutiny of a German soldier whose duty it was to guard, and be, to all intents, hostile.

Part VI
God Is Truth But Man Is Born To Die

    No, we had few opportunities to escape. In the beginning when the chances were best, it was shrugged off immediately as an impossibility. Later, there was too much distance to cover, too many guarded bridges to cross over large rivers, and the physical energy wasn't there to make any plan seem plausible. Yet before leaving the matter, I should mention the one bizarre opportunity to take advantage of the Germans that involved me. It was just another of the endless chain of days when we walked in the cold rain. About midday we came to a village and stopped for rest.

    With others I found myself seated on the sidewalk of the main street with my back propped up against a store front. Above me was a display window. To my back was one of those iron grills which are set at sidewalk level to send light into a basement. In front of us our guard stood leaning on his rifle while smoking a cigarette. I remember hearing no noise from the basement but suddenly I felt a movement behind me. To say I was startled is understatement. Even of curiosity there was none. Whatever was about to happen did not penetrate through the thick layer of ennui that covered me.

    Yet, the movement could not be disregarded since a fumbling hand was reaching for my pocket. My first thought was the guard in front of us, but he was looking away and did not turn. The hand continued to fumble for my pocket and was attempting to put something there. All I knew was that whatever it was I didn't want it. The hand persisted, and finally in a kind of desperation I looked down to my side. There to my great disgust was an automatic pistol which the hand of some unseen person was trying to push into my pocket. Now I was alert and without seeming to move I slipped my hand down to cover the pocket. I wanted that pistol about as much as I wanted a broken arm. The two might easily have been synonymous at that time. Looking at that cold steel I could sense the metal screaming, and once our guard heard its cry he'd hit me in the face with his rifle butt, or worse. The last thing in the universe I wanted then was a weapon. It was a passport to a grave and though the person in the basement couldn't know what I was feeling, he was dangerously stretching his own luck as well as mine. So, a sort of wrestling match ensued. The hand trying to put the pistol in the pocket versus the hand that wanted to keep it out. In the end, I won the struggle when we were ordered to stand and continue the march.

I suppose, looking back, that the intention was good. I have tried many times to think of who that person might have been, and what, eventually, was his or her fate. There are no answers, and while I realize how exciting it would be to weave a fabrication, embellish the small facts and say that I was handed a cryptic note directing me to the Resistance Movement who would help me to escape, it didn't happen that way. Those are the exceptions, the pretty tales that catch the imaginations of the gullible who still

God Is Truth But Man Is Born to Die

"...There to my great disgust was an automatic pistol which the hand of some unseen person was trying to push into my pocket."

believe that war is game of wits and courage, or that medals come to the courageous. Some fools believe the whole bloody show of war is exciting because it is an extension of life when actually it is the cessation of life. One American writer has carried that to the limit by inferring war to the writer is something that is quite irreplaceable and a distinct advantage over writers who have not had the experience. To make his point even finer he says that civil wars are the best since they are the most complete. Of course, this writer has seen war, and later has wet the whistle of his thoughts with notions of a civil war. So, the statements are valid, just as it is valid for him to justify killing by saying that all is well as long as you kill cleanly and avoid the gut shots. Being a hunter, he has done both. He kills clean, and, on occasion, he makes the gut shot. If the one justifies the other, then there is no argument, but it is difficult to accept that it does. So, does a writer who has experienced a war become an expert on killing? Can he say that gut shot men are canceled out by those who were killed cleanly or that his wartime experiences are irreplaceable because he has an experiential advantage over other writers? War, as men experience it, is a complete mess in which the present is a void. It is not what the directing generals, syndicated columnists nor the people who read their dispatches say it is. If war is part of life, then let us all be about the killing for only then shall we find what we are seeking.

Water to the prisoner of war is the very flow of life. In our case, eating only bread and the infrequent bits of other food we might get, water had become a substantial and necessary part of our food. There were times, yes, when we went without water, but never for very long. Our captors knew that it was something we had to have, and each morning and night we were watered like so many camels about to go out into the Sahara. During the long days that we dragged our bodies over the roads of Germany, we would eventually stop for water. We were not deprived of water except in rare cases when, presumably, water was not available. Yet of water, there are haunting memories which tell me that no other substance of the earth could give us the merciless, twisting anxiety in the same way that water did.

In towns and villages, we were given water. In the open countryside we stopped by many small streams to drink. Always and ever this privilege depended upon the men who guarded us. If they were of one evil mind, we went thirsty. If they were human beings, we drank. The human element, or lack thereof, being the determining factor. And yet, because of the heavy rains that kept us wet at least six days out of every seven, we were never driven hard by a parched throat as men might be in other places. However, there were days when the pressure of thirst was strong and compelling. So much so that in one case it resulted in a man's death.

The land through which we walked was the rich farmland of Bavaria. It was richer than ever in those early months of 1945 because of transportation disruptions throughout Germany. As a result, much of the produce had to remain in the district where it had been grown. You could feel the food wealth all around you and see it reflected in the meals that people brought to our guards. Everywhere you could feel the abundance of the land. Cattle were fat, hay barns full, everything looked bountiful, and the people themselves seemed happy, well fed and content. All this despite the horrible war that was then overwhelming the country in which they lived. None of this brought nourishment for us, no matter how many times we might watch our guards eating a steaming plateful of wonderful farm cooking. Neither did it bring envy or self-pity or anything that might be expected. We were so thoroughly conditioned in our life that the sight of real food was merely a phenomenon to be observed.

The war did little to harm or reduce the people of Bavaria. If anything, they were healthier and fatter when the war had ended than when it started. This was the land we had marched through on the day that a man died for water. When walking in column our orders were strict. We did not leave the ranks for any reason. When we stopped to rest, we did not move so much as an inch from the spot where we had fallen. We were marching when it suddenly happened.

It was so sudden that most of the men were unaware until they heard the crack of the German rifle. An Indian soldier in the uniform of the British Colonial Forces had broken from the ranks and went no more than 20 feet to a small stream that ran its course parallel to the road. He bent down on one knee to drink, and his

God Is Truth But Man Is Born to Die

"There was no shout, no warning, just the crack, and the smat of the bullet hitting an unsuspecting soldier bent over a stream."

lips were a fraction of an inch from the cool water when he pitched forward with a splash. His face rested in the stream which ran red from the blood coming out of the hole in his forehead. He had been shot in the back of the head. I cannot forget the look of satisfaction on the face of the tall German soldier who had made the good shot. There was no shout, no warning, just the flash of the rifle being raised to the shoulder, the crack, and the smat of the bullet hitting an unsuspecting soldier bent over a stream.

Late into the afternoon the many Indian soldiers who were with us at that time chanted in Hindustani, "God is truth, but man is born to die, God is truth, but man is born to die." The dead Indian soldier was being given the soul-liberating, flame-destroying burial service of his religion. No, you do not forget such scenes as that afternoon when an Indian was cremated by the side of a country road in southern Germany. I think the Germans were too awed by the whole situation to deny the last rites which the Indian soldiers insisted upon. It was inconvenient to be sure, it meant packing most of us into the corral of a nearby farm. It meant taking several prisoners into the nearest town to bring back the great quantity of lumber that was needed for the funeral pyre. More than anything else it meant that we had to stay in the vicinity until the sun had set. For us this was rest, but for the Germans it was a nervous, restless afternoon. Though they did not have to do any of the work or service, they did have to pass an afternoon with the consciousness of what had happened because of one of them. Had the dead man

God Is Truth But Man Is Born to Die

"*God is truth, but man is born to die'. The dead Indian soldier was being given the soul liberating, flame-destroying burial service of his religion.*"

been a British soldier or American it would never have happened. At most there would have been a shout and perhaps a good clout on the side of the head but nothing more. But no, he had been a colonial, a dark skinned, exotic soldier, and for that he died.

We saw the German mind in action regarding their nationalist groups more than once. There was a strict code to be followed, and as far as I can remember, the Germans never deviated from it. Naturally in our time as prisoners we came into contact with hundreds of German guards to say nothing of the thousands of others with whom we had small, but often meaningful, contact. We found a high consciousness of nationality among them all as it applied to different races of prisoners. In their actions, they would reflect these attitudes towards us in many ways. There were so many cases of persecution or favoritism that it was easy to infer that among the many perversions a dictatorship gives its people, the murder of the Indian soldier was one of the major examples.

At its lowest level it was as innocent as townspeople gathering to watch a prisoner when he squatted over the ground, or stood to make his water. In those moments, the women and children would come from far and wide to study the foreign anatomy. They were most interested in the British and Americans and the least interested in the Poles, Russians, and British colonials, except for Canadians and Australians. Presumably, they felt that the east Europeans were closest to themselves whereas the others were the foreigners who might have something to show in anatomical differences.

Of the Indians or Africans among us there was no interest or, to put it another way, they had the same mark that the Jews had in the Nazis classifications.

Of these "Untermensch", the Germans stood in disgusted awe. Across Germany, however, ethnicity-based persecution was finally being reversed, and the fortunes of war had everything to do with it. The Americans, British, and French were making deep advances into Germany in 1945. It was easy to see that these fanatical National Socialist groups would be the ones that would ultimately reduce Germany to the status of a conquered people. This was so well known that the treatment given to prisoners reflected the pattern of the war and its aftermath. There is no denying that we westerners were among the favored people. The Russians, the Poles, and the colonials on the other hand were either traditional enemies or inferiors, and the Germans believed there was something positive in getting rid of as many of them as they could before the war had ended. They were the ones to bear the burden of persecution, the ones to die if sacrifice had to be made. This came about in many ways, but perhaps the clearest example could be seen in the way that they marched us on the roads. Heading our column were the Poles, the traditional enemy, the bastardized Germans, the traitors to a national cause. Following them were the hated Russians and those who had skin that was not Aryan fair. Following them were all the other scattered nationalities who opposed the Germans, and at the end of all columns were the British and Americans. The reason behind this was simple, and deadly. For almost all the months that we walked, our best chance at dying was from our own people. These were the young, "trigger-happy" airmen. The American and British Air Forces had conquered the air over Germany and were making daily forays into German territory. For most them these missions were great flying parties. There was plenty of ammunition to shoot, and no opposing air force to harass them. Anything was a valid target to them, especially if it moved. Moreso, if it was a column of marching men going down a German road. Prisoners were an ideal target, and although their orders were to the contrary, the Allied fliers seldom bothered to identify the men below. Out of the sun the P-38's or 51's would drop from the sky eager for the kill. And kill they did. Their greatest achievements were always at the head of a marching column.

The law which says the end of the column will stand when the front is down was well known to the Germans and they marched us prisoners accordingly. Naturally as the incidents increased, we began to seek protection. Banners identifying the group would be unrolled when the first aircraft were sighted. This would save us in perhaps fifty percent of the time, for the fair-headed boys of the air often suspected that Germans soldiers were using the banners to protect themselves. So, we'd agreed among ourselves that when planes descended, we would not move. Our calculations were that if we did not run for cover, if we waved and hoped that the flier

God Is Truth But Man Is Born to Die

"*For almost all the months that we walked, our best chance at dying was from our own people. These were the young, "trigger-happy" airmen.*"

would make the initial proper identifications, he would say to himself that these were friends below. We were wrong. In fact, we could not have been more wrong about the men in the air, or ourselves on the ground. Why it was, who knows? The Air Force had plenty of ammunition with which to make noise, and nothing could interfere with the pleasure which came from shooting at large groups of men marching down a road. Meanwhile on the ground it is one thing to say to yourself that you, too, will stand in the face of incoming, hot-shooting aircraft. In practice it never worked. For a time, yes. But after the first few anxious moments the nerves cannot stand it any longer and every instinct says it is time to run. So, the prisoners would run, following the German guards who were always the first to run. When the aircraft pilots saw us running, they were convinced we were the German army, their prime target. You couldn't win no matter what, so we contented ourselves by cursing the men in the air and all their ancestors from Terre Haute, Kansas City, Sacramento, or any of the shires.

In time orders were to get so strict, there were so many prisoners on the backroads of Germany, that the Air Force did try occasionally to identify men on roads. Finally, the situation improved to the point where men no longer gasped out their lives on German roads. Our fliers would come down wings a-flutter, and fly circles above us. They would dive and turn and try to express that they were with us from high in the air. 'Take heart, cheer up lads', and all that. But there was still the ammunition to be fired, so off they would fly down the roads to the next village. There they would put on a fine show shooting up the houses, the women, the children, the cattle, the town fountain and the pissoir if there was one. If they carried a light bomb, things were even better. Back they would come to us for the final wing over salute and affirmation of their devotion to our cause. They would point us toward the village ahead, now full of smoke, fire, screaming wounded, and the quiet dead. Minutes later, we three or four hundred prisoners would come marching into the German village that our proud boys of the air had just destroyed.

There is no more to say of such events told. Men died on the roads. Death came to the villages. The facts are we walked on.

Part VII
"Schweinhundts"

It is night and the rain which has been falling all day continues. Prisoners of war are miserable. They walk so bent over that seemingly their own balance will carry them face forward into the ground. All of the men are inflamed with pneumonia. Heads and shoulders sag, legs are stiff and move automatically. When they do stop to rest each hour, the men collapse like rag dolls right in the middle of the hard pavement. Immediately they are asleep and stay that way until the foot of the guard nudges them or other prisoners pull them to their feet. Many cannot get up without help.

This goes on throughout the cold, soggy day and well into the evening because they have not found a field large enough to hold us for the night. It does not matter. When huddled in a wet field and allowed no fire, men would gather in groups for warmth. With knees to the center and heads fallen forward, they would appear as some grotesque animal. In this way they would suffer through the night, not being able to keep warm or sleep except for the illusion that they dream it to be so. It was black that night when we finally came to a town after a day of stiff-jointed torture. Orders were given to halt, and we dropped to the wet curbstones.

It was late and there was little or no light in the town, and no movement. As usual our guards lighted cigarettes and we all vicariously sniffed the air for a taste of the exhaled tobacco smoke. Behind me, and across the sidewalk, was the door to a house.

"Schweinhundts"

Silently and with- out the tell-tale click of a latch, the door slowly opened. A figure I could smell but not see caught my attention, as any animal would instinctively sense when approached from behind. There was the slightest nudge on the shoulder, and as I turned, my instincts told me that I was about to receive something. I cautiously turned to take whatever was to come. Men on either side of me were alerted by the movement, but they did not move. Our eyes never left the back of the guard who stood in front of us. Then, the mysterious figure slid something into my hand and quickly moved back into the darkened house. I momentarily thought that I was hallucinating, for in my hand was a warm blackberry pie! For minutes it seemed suspended in the air, not a part of me, and I held it without any certainty that it really was resting in my hand. Our guard shuffled in the darkness and seemed to turn but before he could, the pie was safely under my shirt.

At first it was difficult to comprehend, yet the undeniable presence of the pie under my shirt was enough to make me fearful of detection and loss. The darkness helped of course, and when we rose to begin our walking, I slipped the pie from under my shirt and we, meaning myself and the five men who walked closest to me, eagerly began to eat. It is well that we could not see for we dug into the pie with our fingers, tore it apart, dripped juice and had eaten the whole thing before we had gone more than ten steps. That small town remains a mystery to me. There were places like that all along the way. It seems to touch one's sanity to say that in the middle of a wet, cold night in a hostile country you are suddenly handed a warm tin of pastry. But there was another thing that happened in that town that was to leave me perplexed for a long time.

As we walked, we came to a corner and our column turned to follow a new street. Here there were people, the first we had seen in the town. Fifty or sixty had come out of their darkened houses to watch the prisoners walk by. It was an ordinary enough thing, but this night it was late, it was wet, and it was hard to imagine that there would be enough interest to bring the people out of their homes. They stood in a solid group on a street corner and because of the darkness it was impossible to see any faces. When I came to that corner, I passed very close to a group who were standing along the left side street. Just when I was closest to them, I heard a woman speaking. What almost dropped me to the pavement was the fact that she was speaking English and what she said was this, "I do hope your stay here won't be too long." Of the prisoners, no more than three of us heard her for she spoke softly. It was impossible to see the woman or to even attempt an answer. It all happened very fast, and by the time the words had penetrated our fogged brains, we were beyond the corner and there was no turning back. Deep in Germany, far from friends, that woman's kind words came out of the dark night and hit me like a thunderbolt. I can still hear her say, "I do hope your stay here won't be too long."

After you have been a prisoner of war you wonder for a long

"Schweinhundts"

"...I slipped the pie from under my shirt and we, meaning myself and the five men who walked closest to me, eagerly began to eat."

time what it was all about. For weeks and months and years, perhaps forever, you try to piece the torn fabric together. What was real and what was a dream? How much of it was really pain, or, looking back on it, was there any? Many men, perhaps even millions of men who fought in the war, assume the attitude that they have forgotten the experience. I do not believe that they have forgotten the war. I tend to believe they were never aware of it. There is a difference, and it may be one of the clues to why human beings can repeat the folly of war two or three times in a century. Like it or not, the experience leaves its mark and makes its way into the attitudes we call peace. No man goes unaffected no matter how much he shrugs it off.

For myself, I am not much interested in who affirms or denies what. That will come out when the individual interreacts with others. What fascinates me is the process of conversion from man of peace to man of war, from sanity to killing. It's where reality becomes too stern and where the emotions rebel. For it is given to us, as any former prisoner of war can attest, that when the unacceptable shows its face, we can slip into our subconscious minds and create a dream or fantasy that is more to our liking. This gift, if we can call it that, is common to mankind and does not rest only with a wartime experience. Everywhere there is evidence. In all walks of life, we can see it if we look and honestly admit that what we are seeing is reality. This is a stern test of human stuff as everyone knows. And, again as everyone knows this world is

"Schweinhundts"

full of deft escape artists. They are everywhere and wear a million gossamer disguises. Time has not helped to cut through the cardboard wrapping of one event that occurred during those months of captivity. I don't think it ever will. The event was far more real than many things that have happened since. When compared to all the other experiences I'd suffered while a prisoner, I find more reality than unreality in it. If there is anydoubt at all, it is reserved for myself and not for what happened to me. If some events have slipped into mental pigeon-holes that I cannot probe, I will not peer into those holes and claim to see when there is no light with which to see. Our movement was always on foot. Once we rode in trucks for five or six miles. If you were close to death, you were pulled along in oxen drawn wagons. The only other time we ever left the earth was a freezing night when we were high in the hills that lead into the mountains that are the eastern limit of the Austrian Alps.

It was out of the question to think that the prisoners could go into a field and survive the cold hours. The farms in the district were all small and no barns were large enough for the large number of men in the group. We walked on and it was dark before we came to a town which was large enough to have public buildings of any size. For us the guards chose the railroad station, a large, brick, clock-towered monstrosity that had only space to recommend it. Apparently, no trains were expected for the building was deserted except for one worker. Inside it was still very cold and the single large coal stove could do no more than warm about one fiftieth of the area. However, we were out of the cold and the wind-driven rain, which turned into sleet in the hours just after dark, and rampaging snow by midnight. The prisoners had settled down on benches and floor. Most were asleep almost as soon as we entered the station. At midnight with ice forming on windows coated with fine snow, the high lonely wail of an arriving locomotive awoke us to a world outside of our sleeping bodies. Many guards came in and roused us all. They shoved and pushed and gave hard German commands that most of us never understood. But none of us failed to understand what was happening when we found ourselves outside fighting the wind and snow on the platform of the station. We were cold to the bone in a matter of minutes, and we stood in great clouds of swirling snow as the train rushed in and slowly came to a stop. Unresisting, unthinking, unknowing, unbeing, we stood like so many cattle waiting to move. Stamping, blowing, snorting, puffing, we waited on the platform. That train was a mystery in the night.

We did not know where it had come from, where it was going, or even why it was there. Our guards, however, knew more than we did, and they began to open doors and count off six prisoners to each compartment. In alternate compartments one guard went in with the men. Voices shouted, doors creaked open, doors slammed shut, as steam whistles blew in the cold, snowy night. Finally, we were all loaded in, and the train began to move slowly down the

"Schweinhundts"

"The only other time that we ever left the earth was a freezing night when we were high in the hills that lead into the mountains..."

tracks, dusting the world with snow as it gathered speed. It took time to realize what had happened, our reactions were so far behind our perceptions, but we were on the train and moving. I was in a compartment with five other prisoners and no guard. There were two strange people with us, a man and a woman, though it took us some minutes to realize their presence. The compartment was warm, the most comfort we had known in years. There was the plush smell of the seats, of sausages and beer, and the fragrant, though usually unpleasant, heaviness of stale tobacco.

We were still settling into our seats when the woman spoke to the man. "Schweinhundts," she nodded in the direction of the prisoners. It was a single word, and she made a gesture that may have been an imitation of spitting on the floor, though it was too dark to be certain. Then a rapid exchange of words erupted between the two. As much as I could make out, the woman was spilling out all the filthy expressions that had ever been applied to the enemy. The man, less indignant, was trying to quiet her without success. After she had lit a cigarette, she continued her tirade, and was blowing clouds of smoke in our faces. By this time my eyes had become accustomed to the darkness, and I could see that the woman was in the uniform of the German Red Cross. The man who tried vainly to stop the five-minute flow of abuse was dressed in ordinary clothes. The woman finally became exhausted from words and from then until we left the train, she occupied herself with eating chocolate and sandwiches while drinking from a bottle of red wine. All this

"Schweinhundts"

"Schweinhundts," she nodded in the direction of the prisoners...I could see that the woman was in the uniform of the German Red Cross."

while smoking constantly and blowing the smoke into all corners of the compartment. We rode the train for no more than half an hour. We were taken out and marched across a high railroad bridge. The snow was no longer falling, but the air was bitter cold. I gathered from the talk of the guards that the train could proceed no further since the bridge was weakened by recent bombings. Soon we were on a snow-covered road and the night grew increasingly darker as they marched us through heavy forests. How long we proceeded this way is hard to say, several hours no doubt, and it was still some hours before dawn. From time to time, I recall sleeping in the snow on the road and I remember an accumulating heaviness that spoke of being at the outer reaches of weariness. If the guards had not dragged us all to our feet after each rest many of the prisoners would have slept on until they were frozen.

The hours dragged along with the feet, and all sense of time or consciousness was gone. This I know, because from time to time I suffered from a recurring hallucination that I was sitting at the top of a very tall tree and looking down and watching our marching column pass. I could hear the thud of many feet, the coarse breathing, and the sound of the guards shouting up and down the line. I would be there, and I would not be there.

Our road was one that climbed and fell, twisted and turned. Coming around a sudden curve I saw ahead the rooftop outlines of a village. From many of the houses smoke rose in straight lines into the cold, night air. Behind the walls you could feel the warmth of the houses that we were approaching. You could sense people under feather comforters, and you could almost touch the embers of the fires that would cook their breakfast for them in the fast-coming morning. As we came into the village, I saw that it was very small. A handful of houses, some shops, a Square where there was a fountain and a watering trough. Through the village ran a stream and over it at several places were arched brick bridges.

At first this place was just another of the endless towns and villages we were to walk through. Late at night it was quiet, sleeping and nothing demanded attention. I was just about to bring my eyes back to my moving feet when I sensed that there was something unusual about this place that was forcing my numbed senses to react. I looked about the village and immediately understood why I should not march in my usual close-minded fashion. The houses to my left and right were brilliant in the moonlight. Each was colored differently, and all were designed in such a way that made them seem out of some children's book of fairy tales. They were neat and trim and had fences and gates in front of them. Each was a distinct gem with gingerbread shutters and candy canes roof supports as smoke curled above the chimney tops. And, then I knew! The chimney tops were only a few feet above the top of my head. If I raised my arm, it brought it to the level of the upstairs windows where people slept. Fences and gates were only slightly above my knees. All was in scaled proportions.

As we marched along another street in the miniature village we'd come to a row of small shops. The bakery, the butcher, shoes and boots, cloth and clothes. Here, too, the shopfronts were gay colors. The door-tops reached only a little above my waist. There were many items for sale in the miniature shop windows, and all were to scale. The town fountain, the bridges, even the cobbles on the streets were downsized for people such as we had never seen. Here, was challenge and stimulation. Frantically, I looked at bushes and trees and even these had been planted from varieties that would never overwhelm the scale of the village. In desperation, I threw my head from side toside attempting to see all. I had to see all the people, animals, birds, snakes, fishes, anything which would help me regain my wits.

All in an instant I must know why and how all of this was, I had to hear and see, touch, taste and smell. I groped for context and fought with myself to keep moving through this wonderful scene. Information and insight I needed, but there was none for the village slept, animals were out of the weather, and night was master of the way. Smoke from the tiny chimneys rose above the tiny houses of the many colors, cakes and candy. The high steepled church whose clock was above the tips of my outstretched fingers rang out the hour. It was 4 am. As the road descended away from the village and through the woods, the snow decreased and was finally gone.

When the sun rose over the eastern hills, we were once again walking down a country road in southern Germany. After a

"Schweinhundts"

"Each was colored differently, and all were designed in such a way that made them seem out of some children's book of fairy tales."

while I looked ahead and saw the rooftop outlines of another village on the horizon. Faint smoke was curling above the chimney tops. A whistling man passed us on his bicycle and hailed the guards but ignored the column of staggering prisoners.

Part VIII
The Writhing Maw of the Monster

Seldom was the impact of an event as pleasant and challenging as that time when we marched through the Lilliputian village. I have no index to the reactions of others that night, but for myself I know there was a stimulation offered to the imagination and that its acceptance was a warming thing. What it did was to take some of the bitterness out of the long hours and it instilled a warmth that frustrated much of the cold and prolonged weariness.

The memory of that village was to haunt me for a long time. While still a prisoner it remained fixed in my mind and became an excursion I could take whenever I wanted. More than once did I mentally return to that town to look again at the multicolored houses with the short-bread roofs and the candy cane pillars. Often this was better than looking at places of the present or those that might be of the future. From the standpoint of curiosity, I thought of that miniature village for years. I kept wondering what or how it came to be, or why it was there. I suppose what I wanted to know was less important than what I didn't want to know, and I left it in a quiet corner of my mind where I could always find it.

Unlike the village, most of the events that stand out in memory were those when personal participation was part of a larger group participation. They were the times when every man was shocked by some sudden event that impacted all of us. Though a man might be thinking for himself, these were the occasions when he was dragged into the whole unit of thinking. Not often could each man drift off on his own thoughts to study from a distance those things that were impersonal and outside of us. Usually, reaction came because of something that was happening to the prisoners, or because of something that the prisoners were doing. Sights and sounds beyond the sphere of our own small world did not usually excite the imagination enough to pull a man out of his brain fog. Such things as the miniature village were the exception.

Often our guards would stop at farms to have their lunch with a German family. On these occasions we would be seated in the farmyard or a fenced field, locked into a barn, or whatever was convenient. Our guards would then eat in relays. Many of these farms, the buildings, the people, the animals, I can still see clearly. I can still see the incident when hungry prisoners locked in a barn had dug under a thin skin of wall to a storage room for oats. They were busy stuffing their pockets when the dogs started howling, and the guards came rushing with curses and kicks. Then there was a farm where a woman brought out a box of dried apples for the prisoners. No doubt the apples were very good and had been carefully dried behind the large cooking range in a spotless kitchen. But there were about sixty pieces of apple and more than four hundred prisoners. The guard found division impossible and was altogether uneasy, for had he not been armed he knew the starving prisoners would have rushed at his throat. Puzzled and perplexed he poured the apples into a feeding trough for the ten or twelve hogs who rooted around in a pen. Yet, another time the prisoners surreptitiously dug potatoes behind them while sitting on an edge of a field. The guard had been watching their movements the entire time and waited to pounce at the precise moment when the men thought they'd gotten away with it. In that instance, every potato taken from the ground had to be replanted before we left the farm. These and many more were only incidental. Others were more of a lesson and while they are vicious and ugly to look upon, they add their touch to the portrait of humanity and to the rotten depths to which human beings can fall. There is no need to foul the air with most of these stories. Stories that reveal some of the German people as less than low beasts eager to take the pride of the British, French, Russian or American manhood and lay them out as dolts and fiends. Germans who at the time of these actions, were no longer part of this earth, but to some black hell where men are lost to depravity and cruelty.

One instance remains, and will remain, vivid. It speaks for itself, and the speech is not good. It was on one of the few days when we were not soaked by rains from morning until night. For that reason, the prisoners seemed to be in better spirits than usual. There was even sun to make the fields seem more friendly though clouds kept coming and going and with them the sun. At noon we stopped at a farm where the guards would take their lunch. The prisoners were taken out behind one of the barns and there; locked into a large corral or small pasture, were allowed to rest. The guards went off to eat leaving two or three of their number to watch us. Across the fields from a neighboring farm a wagon approached our enclosure and when it had come right up to our fence the driver spoke with the guards. He was a typical German farmer of more than middle age. He explained that he had been planting potatoes when he'd heard that some prisoners were across the fields. His day's planting was done, and his wagon was still loaded with seed potatoes. These, he asked our guards, he would like to give to the prisoners since they would not be planted. Ordinarily the guards would have refused the man but this time I suppose they were softened by the fact that they were looking forward to the well-filled plates of food they would be enjoying in a short while. So, they agreed, and the farmer prepared to distribute his potatoes.

Apparently, the man felt that this was a great moment in human brotherhood for he stood on the seat of his wagon and began to give a speech. Most of us had no interest in what he would say, if there was going to be a potato ration, we wanted the food, not words. The farmer spoke for some time and as he warmed to his subject, he began to get excited. This amused the guards who stood on the side. First, they listened, then they smiled, and by the time the speech was ended they were laughing heartily. With his ending words the farmer was more prepared for a massacre than a giving of food to the hungry. It was plain to see that he had worked himself up just as surely as though he had taken a drug. It was an odd speech to be making to starving men about to receive a gift of food, since more than half of his phrases could have been shouted at a dog who had had just taken a bite out of your leg. By the end of his speech, the farmer was in a frenzy. His hair was wild, his eyes matched those of his horses who had reacted to his words far more than any of the prisoners.

We had learned from experience that food, any kind of food, means nothing to the hungry until it is in the hand. Up to that time you can show it off, talk about its merits, hold it aloft and say, "Can't you taste it?" but it doesn't mean a damn thing. There is no substance to it until the hungry one has it. Then, things are different, and men will fight for the smallest crumb.

This the German farmer knew. It was a curious situation since we had the advantage of not really wanting his potatoes, but he needed us before he could give them away and fulfill his true ulterior motives. Many of the prisoners understood this, but once the potatoes started flying it was soon enough forgotten. Finally, when this demon farmer was deviled, he reached behind him and took two or three of the mealy moldy seed potatoes. Like a juggler at a circus, he tossed these tentatively from hand to hand. The prisoners were on their feet and crowding the fence that separated them from the farmer and his wagon.

They watched the juggling act and started an anticipatory howl. Lips began to move and teeth to chew. The farmer howled with some kind of insane glee and threw the three potatoes high into the air above the prisoners. Hundreds of eyes watched the arc of the potato's flight before hundreds of hands would reach for that point in the air where it would fall. The potatoes would continue to fall amidst shouts and cries that arose from the writhing maw of the monster. It was cruel, and the farmer knew that. Much too cruel to not be repeated. Five or six times he threw just a few potatoes, three or four, as though he was warming up for the big act. The starving men disappointed him no more than starving dogs would have if he were throwing them bones dripping with blood. The stage was set and tense. The farmer leapt into the back of his wagon and began to flail his arms. Potatoes flew toward the mass of howling prisoners. The farmer laughed and roared, and as he threw potatoes he screamed and shook. That farmer threw potatoes until there were no more. Exhausted and laughing, so much that he could hardly hold to the reins of his wagon, the farmer whipped up his team of horses and drove through the fields in the direction of his farmhouse.

Of the prisoners who had become willing marionettes to the strings this fat ass pulled, they had devoured his rotten potatoes, and were still scraping some of the crushed ones off the trodden ground and pushing the pulp into their mouths. Many of the men were bruised and bleeding and it didn't seem to matter. Others were still on the ground where they had fallen and been trampled into unconsciousness. Several had crawled like broken animals to far corners and were whimpering with broken arms or legs. Some of the men, hit too hard in the stomach or elsewhere, were still

The Writhing Maw of the Monster

"*The farmer leapt into the back of his wagon and began to flail his arms. Potatoes flew toward the mass of howling prisoners.*"

vomiting as they held to the fence to steady themselves. Others were just dazed at what had happened or were altogether empty again. Many sat and wondered why our guards beyond the fence were doubled up with laughter. Once the mayhem had subsided, it wasn't long before this sickening affair was forgotten by the men, until the next time when they would be ready to murder their brother for the awful taste of putrefied food. Mothers and fathers who waited to hear of the fate of their lost ones would've been devastated by the sight of them at times like this. I'm sure they would have found it impossible to believe that these boys, these men, these soldiers, would have willingly stuck knives into the backs of their fellow soldiers for food.

As for the farmer, I'll call him a sadist, a good psychologist, a modern. He walks on the same street with all of us every day, he may live in the house next door, he maybe an ex-prisoner of war or for those who find all this hard to believe, he may not really exist at all.

Part IX
That Unforgettable Place

It is our easy way to recall the atrocious acts that man seems so readily capable of. These are acts that we deem evil, but it so happens that it's through these instances I can easily remember my fellow prisoners of war, and probably some of them remember me in the same way. We hardly ever spoke to one another, and even our looks were infrequent. Each being lost, we never tried to give another hope or any insight into the facts of the self that might have given a man some comfort in his own lost state. We were introverted, confused and selfish. It is no wonder that little can be reported that was good during those days, or inside the minds and bodies of those men. It was not a time for sacrifice, it was not a time for awareness, or for any outgoing heart. The days were filled with suffering, from hunger and from cold. The one thing that dragged us all down to a common pitiful level was apathy.

Like a wandering pack of wolves, we travelled together and the only time we took notice of another was when a soldier collapsed to the ground. Then, like the wolves, we were all ready to tear the flesh from his bones if it would be to our advantage. The thing that makes all this more horrible was that it happened directly, and without any buffers in between. You can say that men do the same thing in business, and it will be true, but for the individual involved it will not be the same. In-between there will be disguise and deception, mechanisms of society that hide the black facts of the act and allow the guilty individual to come out of the fray with that modern misnomer-a clear conscience. With us that was not possible since our level was low and there were no artificially created barriers to hide behind. That is what brought sorrow and disgust into our world, and we could not escape the voice of our conscience no matter how we tried. What happened happened, and you were forced to see it for just what it was. There were no rose-colored glasses to wear, no bars, churches or banks in which to hide. No differences marked the men, all were much the same, equally capable of good and bad. Where the possible reward was survival, it is not surprising that what came out was poison.

All men were thieves and that was a pity since there was really nothing to steal. Of course, that is not an exact statement for there were things that could be stolen, and, looking at the comparative values, many of the things we had were priceless. Take for instance the simple tin can. Few of the prisoners had such an article. Those who did possess a tin can guarded it as though it were made from gold. Its only use was to drink water from, and even that convenience was small. Small or not, a tin can held the value of a Cadillac automobile to the average prisoner. It was an article in which to take pride, and the use was almost secondary to the effect that the object had on others. Men would plot for days to steal another's tin can and devote their whole thinking process to the act. We had nothing to cut, but anything with a cutting surface was another precious item that became fair game for the pack of thieves. Head covering was another and that might be anything from a beret to a worn felt hat or a woven mat of rags. All of these, and other things just as useless, were sought after. It seems strange to say it, but a man might go to sleep and wake to find that his trousers had been cut off at the knees or that two matches which he had found somewhere had been taken from an inner pocket. More cleverness was displayed in what these thieves could accomplish than has ever been recorded of the organized crime in cities. Perhaps the rising crime waves that always seem to follow a war are due to the fact that millions of men have learned the use of guns and pseudo-toughness, yet it would probably be safe to say that many prisoners learned the art of deception and thievery so well that they have contributed to the post war statistics. Whatever; it happened constantly and may have given zest to a dull life.

Like most I had nothing to lose, though the time did come when I was given a beret, and from then on it was a case of the hunters and the hunted. Still, I never lost the beret for I watched it all of the time and never slept without putting it where it couldn't be taken. That beret is a story in itself. One rainy day we passed some French laborers who were working in a German town. They were prisoners also, but they had been taken into Germany for hard labor. On this occasion, they were cleaning streets. As we walked by, I guess I must have looked miserably wet for one of the Frenchmen made a common joke about this. I replied with what is the answer to that common joke and the Frenchman laughed and threw me the beret which he had taken from his own head. Our guard saw this and since he was one of the rough and tough boys, he chose to act. Many of the guards we had known might have let it slip by. Not this one. He turned to me and shouted but made no move. Then, he ran at the Frenchman and lifting his rifle he struck him with the butt on the side of the head. I remember seeing the Frenchman's eyes go wide and the deep gash on his head spurt blood as he spun two times and fell to the wet pavement unconscious. The guard stood above him. "Ya…gut!", he said and turned back to the column which continued to move down the wet street. To me nothing happened. The guard had had his satisfaction.

To those who looked at the prisoners as they marched by there must have been some amusement since there could seldom have been any sympathy. Dirt was caked upon us for we never washed, and every man had a matted beard that was a jungle growth suited for lice. Our clothes were torn and filthy and after a time it was impossible to guess what nationality a man was from what he wore.

As weeks wore on and dirt found its way everywhere, each man would eye with envy the clean jacket of another and before long there was a general trading of clothes There wasn't much difference between the dirt on one man's torn shirt and the dirt on another. To the eye of the prisoner, however, every man was somehow cleaner than himself. It became a game of trade and all participated. For a Polish flyer's jacket, a man would trade a pair of British pants.

That Unforgettable Place

"He ran at the Frenchman and lifting his rifle struck him with the butt on the side of his head. I remember seeing the Frenchman's eyes go wide."

For a Russian shirt, an American sweater. Before long everyone had worn everyone else's clothes. Naturally at any given moment every man was satisfied that he had done the best bargaining and had the cleanest clothes of the lot. Even this conviction changed when a new group of men would join us wearing cleaner clothes than we had. The trading would start immediately and continue until all were satisfied once again that they had the cleanest. Shoes alone were never traded for there was nothing attractive in any. Whatever were on a men's feet didn't matter for every shoe was worn down to nothing. Some had been given wooden sabots when there was nothing left of a leather boot. Others had blistered their feet so often that they could not wear shoes and for them cardboard soles wrapped round with rags were given by the Germans. My own army boots had been cut repeatedly for comfort and for air and had taken on the appearance of Grecian sandals. When, ultimately, we were no longer prisoners, I was wearing four articles of clothing other than my shoes which covered feet that had been without socks for four months. These four bits of tattered clothing had come from three continents and represented the work of people scattered more than eight thousand miles apart.

None of us had much reason to recall the faces of individual Germans. Large gatherings, yes, for they came out to watch us pass by almost every day. But no single face stood out, no reason made an individual distinctive. Our most memorable encounters with German citizens came when they were in groups and could hide their animosity in the anonymous depths of the herd. This would not apply to the guards. We had much more contact with them and there were reasons to remember one or another of the many guards who led us through Germany in those long months. Out of the thousands of German townsfolk and villagers we passed I can remember only two of them. They were both old men, both very different, and their actions bore out that fact. One of the men was to give elation through his seriousness in a comic situation, the other by his sincerity in a tragic situation. Again, there was the great division in the scale of human action and emotion. Perhaps because I have separated them so widely, I remember them. The hundreds of Germans that would fall in between these two are remembered but not with faces and thus I cannot say that people were involved.

We had come to that unforgettable place and situation where Americans were walking through a town that had been bombed heavily and recently by American planes. This time the damage had been great. The whole town was sighing with its hurt and many of the wounded and the dead had not yet been found. Our guards were tense when we came to the place because now, again, they would be guarding us from their own people, and this was hard for them to do. Unlike other times, not many people had gathered to watch us pass, and those that had were in a tight group that swayed silently from side to side. It was ominous and strange; heretofore the cry of vengeance had always come with the first sight of us.

These people said not a word, only their eyes spoke of their hurt and hostility. From out of the crowd one man stepped and he stood as close to us as he could, He was old though straight. He stood proudly and calmly as he gave us normal greeting. Then he stepped back into the crowd and returned to our sight with a little girl of perhaps eight years in his arms. She was dead. He put her down where we could all see as we passed. That time I was toward the rear of the column, and I could see the old man for a long time. He stood by the dead child and said nothing though his lips quivered, his small goatee moved, and you could see he was holding in emotions that burned deep within him. With all the dignity left within; he tried his utmost to hold himself in control. Inevitably, the moment came when he could not contain his grief any longer. He began to cry, and the tears fell down his face. He would not lower his head or hide his wonderful proud eyes. When most of the prisoners had passed, he broke utterly but his only protest in action was to lift a silver headed cane and strike through the air as he sobbed. Each of us passing must have felt the sting of his blows that would have fallen on us except that he would not release the final humiliating passion.

As the days pass, we continue to trudge down the German backroads that lead to nowhere. Ironically, the steady, monotonous rhythm of the march had the beneficial effect of reducing panic. Experiences such as those of the old man and the dead girl would leave the days ahead emptier than ever. In time, the weight of

That Unforgettable Place

"When the prisoners had passed, he broke utterly, but his only protest in action was to lift a silver headed cane and strike through the air."

sorrow and recrimination settled deep within each of us. Still men must survive, nature must balance, and somewhere along the wet roads there must come a release. In fact, a few days following the horrifying death of that child we came upon such a situation. It was ridiculous and unimportant, but it disrupted our miserable thoughts, and allowed us to move once again in an area where the air was lighter. It was also very important that this be pointed out for us by another human being, and that is where we come to the second of the old men.

It was in the open countryside where farms and villages were small. The road had been straight for hours as it went through a valley. It stretched for miles behind and ahead of us. If the eyes had been keen, they would have detected a tiny black spot in the landscape some miles ahead; a man stood by the side of the road waiting for us to pass. Somehow, he knew that we were coming along, and he waited patiently in the rain. When we approached none of us could fail to look at him, for he was the only strange person for many miles around.

He was an old man, well dressed, dignified and a caricature of something or other as he stood there in the rain beneath a huge black umbrella. He wore shiny black leather boots, a waterproof coat and a cloud of smoke curled from under the edges of his umbrella. He had a tiny, precise mustache with waxed ends that turned sharply upward as though to draw attention to his lively black eyes. A smile, beatific enough for that

"As the men marched by, he continued in his German song, and counting precisely, he handed every fifth man a long black cigar."

of a saint, made the whole lower half of his face seem younger than his years allowed. To complete the strange picture, he held in his hands a book which he had been reading while waiting for us to approach him. He could not have looked more comfortable and content if he had been sitting in a deep leather armchair before a warm fire.

The rain was falling heavily, the road and the fields were very wet. Our gentleman was growing larger as the distance diminished all of the time, was dignified and his composure was full of mystery. Finally, when we were just about to arrive at his waiting station, he closed his book and put it away in a pocket of his coat. From somewhere inside his coat, he removed two wooden boxes. With a silver penknife he cut the seals on the boxes and pushed back the lid of one. It was at this point that the first of the prisoners began to pass him. He stood very straight looking into each face and after having disposed of his glowing cigar, he began singing in a rich, though shaking, baritone voice. As the men marched by, he continued in his German song, and counting precisely, he handed every fifth man a long black cigar. He continued this as the column went by taking his cigars from the open box until it was empty and then from the second. His song never stopped nor did his eyes ever fail to check off the face of each man that passed.

When the last man had gone by, he fell in behind us and followed for a short distance down the road then he turned off and walked slowly through the rain up the driveway to a farm. The prisoners were left with amazement and cigars. Soon the whole column of men walked underneath a haze of smoke as the precious tobacco was passed from one man to another each having a puff or two since there were fifteen men to each cigar. Our guards, surprised, and in this instance kind, allowed this deviation from our empty routine. Each of them joined us, puffing on a long black cigar, for they too had accepted the gift. Recollection tells me that this was the single instance when the men had tobacco. Our guards of course smoked, I recall their cigarettes as being an unknown brand called, 'Waldorf Astoria No. 5', but their ration was so strict that even when we did meet an amiable guard who would toss his cigarette end to the prisoners there was little to smoke. Two men might get a puff, the last one finishing with hot ashes in his fingers.

Yet even these ashes were carefully saved, for one man among us had a pipe which occasionally would get one quarter filled with ash and could then be smoked for a few seconds. In the first days of capture the lack of tobacco was hard on the men. All had been under the pressure of war and there were few who hadn't been smoking at least two packs of cigarettes every day. Suddenly that stopped and the chemical processes of the body naturally rebelled. To smoke the men would try anything in the way of leaves or grasses or tree bark that they could ignite. They did get some smoke, but they never got the taste they were seeking. Later, when starvation had numbed all the senses, it didn't seem to matter much. Tobacco was gone but never really forgotten.

Part X
Like Dante's Inferno

That spring as we traveled almost five hundred miles through Germany, we hardly ever had what would be called a night's sleep. What we did get was broken in small pieces by cramped nerves that had to rebel at what was being asked of them. Most of the time, when we did get a few unbroken hours of sleep, it was an uncompromising request that the body had to answer if life was to go on. There was no replenishment of energy in any of it. It met the barest of minimum requirements and with that we had to be satisfied, had to carry on with the never-ending walking. It is not surprising that I am unable to think of sleep in that time except as a cessation of walking.

Few nights separate themselves from all the others even though there were times when we did get what might be termed a satisfactory sleep. Nor is it easy to single out which nights were worse than all the others. Ours was a life that accepted the open wet fields of the countryside for a bed. Having accepted that we had to take what went with it. For us, due to the time of year, it was wetness that you could not ignore and coldness that mattered only because it made the wetness even more miserable.

It was on one of the wettest of nights that we were fortunate. Some kindly fate had smiled and said tonight we would have a roof. If it had been fate, then fate had chosen wisely, for it came at a time when the prisoners were just about ready to stop swimming against the inexorable tide. I think every man among us knew a sickness which in other times and places would have seen us all in a hospital bed. An arbitrary limit had been reached by many of the prisoners, and another night in the elements would have been physically catastrophic for quite a few of them. Our guards were sensitive to this but that was not enough reason to keep us from a wet field. What gave us a roof that night was a rampaging stream and a broken-down bridge. For this reason, we had to detour on a country lane to cross the river at another point.

During this inconvenient detour, the storm that we'd been in all day turned from anger into monstrous wrath. Something alive in the atmosphere had warned men that this was not a night to stay out. The shrieking, howling wind challenged all doubters. Here, we prisoners had found a friend in the natural voice of an informed wind. Soon we crossed the river, and in the darkness ahead of us we saw the outlines of a high and unnaturally large barn. I am sure that many of our guards did not know what kind of a building this was, nor did the prisoners. It was higher than any familiar barn, it was shaped in a different way, and most curious of all, it did not have solid sides. In place of straight boards there were laths, and these were so hung that they could be adjusted like jalousies or venetian blinds.

Apparently, some structural defect had condemned the barn many years before, and you could sense that the building was not in use. Add to its abandoned aspect the fact that loose timbers creaked in the wind, that the thousands of shutters which made up its sides all moved with a noise like the swishing of bats, that large doors swung back and forth with great crashes, and you will see why the Germans and their prisoners hesitated.

But the night was insistent, the wind was hard driven and biting, so we headed for the shelter of the barn. The barn was enormous and sent us no welcome. From the moment we entered, it became an unwilling host to three or four hundred men. We were like invaders and the timbers of the barn shook as though to throw us all back into the black night. On the ground floor there was old, rotten hay. This was the home of a civilization of rats. They probably had us outnumbered ten to one, but we gave that no thought. For us this was shelter and the men made a wild scramble to secure some place where there was hay, and if possible, a corner to be out of the way. The ground floor was hardly enough so our guards beamed their torches to the higher levels of the temple-like barn which seemed to stretch high and into the sky. Above there were only pole rafters, no flooring at all, and in between these poles were hundreds of smaller poles. To the uninitiated the effect was a maze, and all was made hideous by the dim torches which tried to pierce high into the darkness.

Still, it was somewhere to go for the ground was now covered by unthinking, uncaring bodies who had taken all the space and were being trampled down into the straw by those who continued frantically to find space. The men began to climb. First to one level where they would take up all the space on the pole rafters which were also walkways. Then to the second level and on to the third and the fourth. In effect, it was like a macabre illustration for the story of Danté's Inferno, and the voices of the men, the shouts of the guards, the squeals of the frightened rats, the wing whooshes of the birds and bats added final touches to the bedlam. High on the rafters the prisoners settled themselves on twelve-inch-wide boards, dangled like monkeys in a prison ship cage, disturbed the spider and bug inhabitants of the abandoned tobacco drying barn, fought off fear, hoped for balance, then cursed the night and searched for sleep. It took hours but eventually there was silence of a kind, but it was like no known silence. Meanwhile, the howl of the fierce night continued to damn the hours.

Sleep under such circumstances is like being drugged and the first moment of awakening has in it all of the taste of gall. In the morning the prisoners were all stirring early. Outside the skies had cleared and a new day, perhaps a better day was about to begin. The tobacco barn by the light of day was even more of a shambles than it had appeared the night before. Everywhere there was the litter and decay that comes to a building long unused. The prisoners who had climbed high in the night now came down to the ground and there was a constant aimless milling around the barn. The guards had gone outside and had barricaded us in with heavy timbers wedged against the doors. Some of them had gone off to seek our daily bread, others were seated around a fire they had started while waiting for the day to warm. Inside it was not pleasant. The ground floor was a solid mass of standing men, all shuffling against the morning cold, all heavy eyed and bad tempered from the monster-sleep of the night before. We waited thus for more than two hours inside the malodorous barn, and none waited patiently for there would be no reward when our time was up. We could expect our daily black bread, some some water and another departure down a long, long road.

It was as we waited that a distant sound like the sighing of the wind was heard. At first there was little significance in the sound. Like a far-off stream, it was a steady quiet flow that seemed to be moving closer. The men became curious for they could not say what the sound was. Soon, all the prisoners were alerted to the strange noises that were close and coming closer. The men had crowded to the walls to see what was about to appear through the cracks of the shutters. Before the source of the sound was seen, it was defined. People were approaching, many people.

Yes, a crowd was approaching the barn and we were curious, for somewhere inside each of us was the eternal hope that the presence of people would bring some good our way. The walls of the old building seemed to bulge out from the pressure of the men inside, there was silence within, the hushed air that signified men were waiting. It wasn't long before people came into view over the rise of a small hill. There were many people, and they were mostly women with a few young children and older men scattered among them. Instead of taking the easy pace of walking, they came running

Like Dante's Inferno

"High on the rafters the prisoners settled themselves on twelve-inch wide boards, dangled like monkeys in a prison ship cage,…"

over the hills like Teutonic knights rushing across the ice to smash the men of Alexander Nevsky. Their intended destination was the building in which we were locked. In the early morning light, they came swarming through the fields, close to one hundred of them. Clustered together, they made a tremendous amount of noise for all were shouting and puffing and laughing. They were all country folk who seemingly had stepped out of a Bruegel canvas, as the colorful skirts of the women billowed in much the same way. If all the younger men of Germany had not been away at the war, you could easily envision these women in a different scene. It would be in the same light of dawning day, it would be warmer, the fields would have golden hay, and all would be dancing and singing or rolling on the ground with their men. This time things were not so gay. With an unknown purpose to fulfill, the people came running toward the old ramshackle barn which was filled like the deep hold of a ship carrying emigrants to a new land. We waited, hushed, and the men continued to press to the walls.

Picture this scene if you will. There were more men than could possibly see through the shutters of the wall. To correct this difficulty the prisoners climbed over the heads of their fellows, they sat on their shoulders, they clung to the walls like spiders and monkeys, they all fought for a chance to touch the view. Meanwhile, the people had arrived outside and began a traditional dance around the barn that had all the primitive excitement of a tribal war dance in Africa. The women sang and danced, the old men

Like Dante's Inferno

"*The women danced as temptresses. They danced with glee, scratched their bellies, bared their breasts, offered food to a hundred stretching hands.*"

hobbled and hollered, the children imitated them both. Inside the prisoners were a seething mass of frantic bodies. Eyes could not comprehend what was happening outside. The women flourished their skirts, tossed them high over their heads, jumped up and down until the heavy wool stockings rolled down their calves, and all the while they sang demonically. The tobacco barn shuddered from their noise and rumbled deeply from the howling of the men inside. This was the madness of a fantastic orgy, the time when human beings slithered from their skins and made a run for the jungle. It continued and soon the women were pulling apples and breads and sausages and cakes from between their bouncing breasts.

This was the signal for prisoners to go mad, and they did. I will never know what could be seen from the outside, and the greatest understanding would have come if there had been the ability to turn this nightmare inside out. I can only visualize what it looked like. The women danced as temptresses. They came as close to the building as they could without touching the mass of hands that had been thrust through the slits in the wells. These hands, the hands of four hundred prisoners, must have had the effect of snakes for they tossed and stretched until the muscles could extend no more.

This excited the women even more and made their temptations more eager and daring. They held to their food tightly and brought it to where fingers could touch but not hold. They laughed and rolled their hips; they threw apples back and forth to each other. They danced with glee and scratched their bellies, bared their breasts, offered food to a hundred stretching hands and danced away from the touch with the nimbleness of nymphs. What more is there to say or how to say it?

In time the women were spent and could not laugh or tempt any any longer, the old men were worn down and the children snickered with the ending of delight. Homeward they turned with the same fatigue of having just finished seeding a large field. Their song was subdued, their food was back in their skirts or blouses, their bodies were spent. The hands of the prisoners, thrust through the walls, continued to move in spasms that made the fingers curve without control. It was not until the last of the people were out of sight that the last hand was withdrawn from the wall. In a heap, the men had collapsed, body upon body they pressed down the straw and all that it contained, the bodies of dead rats, mice and lice, blood and futility.

Part XI
Coda

Who of those among us could bear the burden of these hardships or hope to put all the puzzling pieces together in a way that there might be a small semblance of meaning? Not the cold and hungry prisoners who are without hope and ceaselessly dismayed. Not the changing guards who walked with us and waited for their relief to meet them at the next town. Though their days were better than ours, they were still soldiers, and theirs was duty. I'm certain that there were times when they could feel our anguish. After all, they also were prisoners of a kind, and the relentless days of war made us all slaves to the same pernicious condition. It is too much to believe that the people of Germany could have shifted the weight. They were held in the same vise of conditions that had made them accept the facts of men killing men, objects destroying objects, and faith feeding like a vulture on the innards of the ideal.

For people across the oceans who went through the war unscathed, with their homes intact, and no privations felt, they'd participated with their hearts for what little that could mean. Nowhere could the burden of these faults find rest except here, there, and everywhere men and women had understanding. They were the ones who really suffered. Not the prisoners or the dancing women. Certainly not the elite guard of the German army or the Allied High Command. Only those who understood the movements of human beings, that a citizen's responsibility was more than the call of patriotism, and that war touched everything related to life, death, war and peace.

Yet nowhere did we sense familiarity. Here was a foreign land, an enemy people and the fact that every man who was a prisoner was doomed to know nothing beyond himself. At long last, it was in one of the larger cities of southern Germany that the clock of our time was to stop. When we arrived, we saw once more the devastation that had been brought to Germany from the air. We were familiar with the sight of railroad yards reduced to rubble, homes, apartments and stores left shattered or as skeletal ghosts, streets torn up and trees made jagged. Here was the quiet of desertion, complete to the lack of a single bird who might have given song.

We had come to another of the many towns and this one is remembered for the smaller details rather than the large. Take the instance of a certain six-story apartment house. This was noted because it had crumbled onto itself and yet had kept its facade. Standing next to it was another where the reverse had happened. It was still more or less intact except that its entire front had been blown off and exposed its many rooms. They were one above the other like so many holes in a beehive. High on the fifth floor of the apartment, a window box of bright red geraniums was in bloom. Beyond the flowers was a bright blue sky framed by what had once been its window. I suppose if the intact facade could be moved over to the next building and somehow attached, we might have an intact building once again, but then there would not have been red flowers floating against a blue sky. It may seem strange but when my eye saw those flowers, I somehow knew that the war and my days as a prisoner would soon end. To each prisoner there came a time when this happened, when there was some signal given, and the knowledge of the end was firm and clear. Through the hard and bitter months, we had walked more than five hundred aimless miles. Spring was upon the land but not in the heart of any prisoner. Even at the end, hope was gone. We all knew that either the war must end, or we should. Fortunately, it was the war that would end and with it our liberation. The events at the end moved swiftly. Five thousand tortured prisoners were held in a city now under siege. In camps, in tents or on the open ground they were guarded by an army of soldiers who were being pressed hard and who were about to flee. Written in the faces of these pursued soldiers was the knowledge that soon they were to take our places as prisoners. In those last days it was our good luck that they did not break under this pressure and bring the weight of their vengeance to bear upon us.

Of the five thousand prisoners more than four thousand had left the burning town with the German army. For those of us who were left behind, the hours of being a prisoner were growing shorter with each series of artillery shells that fell on the town. Two days later resistance eased, and we were finally free. Our months of captivity had ended, and our decrepit physical conditions

CODA

"High on the fifth floor of the apartment, a window box of bright red geraniums was in bloom. Beyond the flowers was a bright blue sky..."

slowly, but surely gave way to health. This wouldn't come from an easy convalescence, instead it will take years after we'd all gone our separate ways and are made to answer in our consciences the many questions we had ignored for so long, or, as even now, to ever seek the answers to man's complexities.

END

The Post-War Years

The Post-War Years - by Scott MacGregor

"Hugh O'Neill is another poet...", wrote American novelist Henry Miller in his book, "Big Sur and the Oranges of Hieronymous Bosch". There's little doubt that Hugh's poetic soul had helped him survive his days of war, and though he'd made it home alive, peacetime would present him with challenges that he wasn't ready for. In 1945, there weren't effective diagnoses or treatments available for veterans suffering from what we now call PTSD. Like many veterans who'd suffered mental trauma, once Hugh's physical wounds had healed, he was literally on his own to heal the other wounds that few could recognize with their eyes or understand in their minds.

Upon his discharge from the Army, Hugh moved to California where he resumed his studies at UC-Berkeley on the G.I. Bill. That same year, he'd unwisely jumped into his first marriage. His wife and others who'd known Hugh before the War were alarmed by his uncharacteristic behaviors which included heavy drinking and physical brawling. Within a year the marriage was ruined by credible accusations of infidelity on both sides. By the late summer of 1947, the marriage had collapsed, but not before their daughter, Seon O'Neill was born. Sadly, Hugh chose to not take a role in his daughter's life. Confronted with a situation he couldn't cope with, Hugh abruptly deserted his wife and child in Berkeley and resurfaced a few days later at Henry Miller's then-notorious bohemian colony on the coast of Big Sur, California.

The Post-War Years

Hugh O'Neill at Big Sur in the late 1940's

This sad turn of events had shocked the sensibilities of his devoutly Catholic mother and Hugh's other relatives in the midwest. It led to long estrangements. Ironically, Hugh's impulsive (many said irresponsible) sojourn to Big Sur bohemia actually changed the trajectory of his life for the better. There he found himself surrounded by fellow anarchists of the mind, conscientious objectors, poets, writers, and artists of every stripe imaginable. Even an occasional movie star would show up. Just like Hugh, many of these seekers were acolytes of the controversial writer, Henry Miller. His revolutionary 1934 novel, "Tropic of Cancer" was declared obscene upon its release and banned nearly everywhere in the world. It remained that way in the U.S. until 1961 when Miller won a historic First Amendment battle in the Supreme Court.

Hugh was just one of many creatives and intellectuals who'd regarded the well-read Miller as their guru, a life coach, as a sounding board for their own endeavors, and, as it was in Hugh's case, the father figure most of them never had in life. The mantra Miller consistently espoused then and until the end of his life was to spurn the 9 to 5 workaday rat race and the "air-conditioned nightmare" he believed the United States of America had become. He railed against America's poor treatment of its artists and instructed anyone who would listen to always "live in the moment", and to regard the tenants of capitalism as the under-pinnings of all wars.

Very little car traffic was coming down the south coast in those days after the War. This meant that Big Sur's stout and hardy com-

The Beautiful and Rugged Coastline of Big Sur, California

Anderson Creek in the 1950's. Henry Miller once lived at the shack in the left background. Hugh had once lived in the larger shack next to the bridge.

munity had the rugged, coastal wilderness largely to themselves. Living in dilapidated coastal shacks and in Big Sur's mountain areas, the bohemians grew vegetables, pulled abalone from the shallows, had endless cookouts on Big Sur's many beaches and coves, and relaxed in the geo-thermal mineral baths at Slates/Murphy's Hot Springs (today the Esalen Institute). They had a blast!

The creative collective that erupted in post-war Big Sur had few parallels in the U.S.A. Up and down the Sur coast and into its high-country areas, dozens of typewriters were clickety-clacking out novels, short stories and poems as world-class creators sculpted, painted, and composed. If an actual artist's utopia was possible, the bohemian community at Big Sur from 1944-1950 may have been one of its closest, albeit highly flawed approximations. Hugh had made life-long friends there and often returned to the Sur coast until the end of his life.

Hugh would have two formal marriages, and at least two unions that were considered common-law marriages. None lasted more than 10 years. At Big Sur, Hugh had met a woman living at the colony who became one of the great loves of his life, Margaret Neiman. Though the relationship lasted only a few years, Margaret was a godsend to him. A warm and caring person, she was also a beautiful and gifted multi-media artist and model who was nine years his senior. Margaret was instrumental in helping Hugh find the peacetime equilibrium he'd been seeking. Henry Miller

The Post-War Years

Hugh O'Neill and Margaret Neiman at Big Sur-1949

had taken note of his transformation as follows:

"Until Big Sur, Hugh had never done anything with his hands. Suddenly, he discovered he could do all manner of work. He made fireplaces for his neighbors, (some worked-some didn't)...he maintained an enormous patch of vegetables sufficient to feed the entire colony. He took to fishing and hunting. He made pottery, he painted pictures...never have I seen a poet blossom into such a useful creature as did Hugh O'Neill."

During the 1950's, Hugh returned to the world of capitalism and to his chosen vocation; journalism. He'd interned or worked for New York and Bay area newspapers before moving to Hawaii where his newspaper career took off. He initially worked as a news reporter for the Hawaii Tribune-Herald, and later as a senior reporter for the Honolulu Star-Bulletin. Between those jobs he'd been a public information officer for the Mayor of Honolulu and the Dept. of Defense. While at the Star-Bulletin, he'd met and married Miné Sato who'd worked as a proofreader at the newspaper. The marriage lasted for ten years and dissolved in Seattle where Hugh had been working as an editor for the Seattle Times.

After his marriage to Miné had ended, the direction of Hugh's life suddenly changed. He left Seattle and joined a Buddhist sect in Oregon. He spent lonely days and nights as a firewatcher and later a cook for lumber camps in the Sierra Mountains. After a 25-year self-imposed exile from his Ohio home, he surprised his long-lost relatives in Cleveland with recurring visits that lasted for months.

His last newspaper job was as an editor for the San Jose Mercury before quitting the occupation entirely. By the late 1970s, Hugh was living strictly in the moment again, just as Henry Miller had taught him. He traveled relentlessly from place to place, friend to friend, relative to relative. He lived in Ireland for a while. He'd stayed at the Buddhist retreat in Oregon. He never had a plan. Looking back, I believe these were all clear signs that the war was still inside him, still confounding him, still bedeviling him.

By the turn of the millennium, most of Hugh's friends, lovers, and champions had passed on. In 2001, Hugh was living alone in Port Townsend, Washington. One day, he collapsed and was taken to the VA Hospital in Seattle. Soon after arriving, Hugh whispered to the attending physician, "I've never felt so tired". Then he turned on his side and passed away peacefully at the age of 80. That day was September 28, 2001. Following Hugh's death, my two sisters and I traveled to Washington State to clean out the apartment he'd been living in. Once collected and sorted, Hugh's life possessions barely occupied the surface of a long card table. My sisters distributed these meager vestiges to Hugh's neighbors who could use them. In accordance with Hugh's wishes, his ashes were given to the Lama of his Buddhist sect in Oregon.

Before leaving Port Townsend, I'd been casting a cynical eye on the run-down nature of the apartment building that was Hugh's last home. I couldn't understand why he chose to live there. Like the building itself, even its inhabitants all seemed to be down on their luck. Then I sat down and peered out Hugh's window. Through it

I saw a perfectly framed view of Mt. Rainier across the beautiful Puget Sound. It was then that I realized why Hugh had chosen that apartment. It may have been a run-down place, but it had a million-dollar view. There's no doubt that's is why he chose to live there. He'd spent most of his life defying convention and anything else that people would refer to as "a normal life". Had he never fought in war, one can only ponder how different his life would've been. The last lines of this story belong to Hugh:

Excerpted from, *"To Those Away"* by Hugh O'Neill

When you walk

where you once have walked

time is an unsubstantial answer

to the fact of your being gone,

and certainty is in the waters

washing against the western shores.

As it is certain

That they come from where you are.

In thinking then of us

Of whom you are so vitally a part

Hear the cry of the shore bird

Caught in his coastal flight.

Know that these home hills

And subtle winds

Will bring you back

Wherever you are.

The Post-War Years

Hugh O'Neill's WWII Ephemera

The Cleveland Plain Dealer
April 6, 1945

(Above) Various items that Hugh wore on his uniform, along with his Purple Heart. **(Below)** News of Hugh's MIA status finally reaches his hometown. A full month had passed before the Army determined he was missing. Hugh was liberated in May 1945.

BIBLIOGRAPHY

The following sources were referenced in the creation of Scott MacGregor's commentary:

Nichols, Lester M., *IMPACT-The Battle Story of the Tenth Armored Division* (Bradbury,Sayles, O'Neill Co. Publishers -1954)

Miller, Henry, *Big Sur and the Oranges of Hieronymus Bosch* (New Directions Books - 1957)

The Cleveland Plain Dealer Newspaper, 04/06/1945 Edition

O'Neill Jr., Hugh A., *Big Sur Poems* - 1953 (unpublished)

ABOUT THE AUTHOR, ILLUSTRATOR, AND EDITOR

Hugh A. O'Neill, Jr. was born and raised in Cleveland, Ohio. After the War he graduated from UC-Berkeley and worked as a journalist and editor for several newspapers including, The San Jose Mercury, The Seattle Times, and the The Honolulu Star-Bulletin. He also worked as a public information official for the Mayor of Honolulu and the Dept. of Defense in Washington, D.C. O'Neill died of natural causes at the age of 80 in Port Townsend, Washington on September 28, 2001.

Gary Dumm is a life-long Cleveland, Ohio resident and prolific artist who has many hundreds of illustration credits in his CV, most notably is his extensive work with Harvey Pekar on *American Splendor*. Dumm's illustrations have also appeared in the New York Times, and La Monde. He also illustrated the 2020 graphic novel, "Fire On The Water", which is a work he'd collaborated on with author, Scott MacGregor. Today, he remains one of Cleveland's busiest illustrators.

Scott MacGregor is a life-long Cleveland, Ohio resident who has been writing and publishing graphic stories for several decades. He is the author of "Fire On the Water", (AbramsComicArts-2020) a critically acclaimed graphic novel about a major Cleveland industrial accident in 1916. MacGregor is also a photographer of Ireland and his photographs have adorned the articles and covers of several magazines and periodicals.

www.ingramcontent.com/pod-product-compliance
Lightning Source LLC
Chambersburg PA
CBHW041436060526
44119CB00106B/442/J